Ten-Minute Real World Science

by Murray Suid and Scott McMorrow

illustrated by Philip Chalk

> This book is for
> Dag Roland

Publisher: Roberta Suid
Editor: Hawkeye McMorrow
Design & Production: Scott McMorrow

Other Monday Morning publications by the author: *Book Factory, For the Love of Research, How to Be an Inventor, How to Be President of the U.S.A., Picture Book Factory, Report Factory, Storybooks Teach Writing, Ten-Minute Grammar Grabbers, Ten-Minute Editing Skill Builders, Ten-Minute Real World Science, Ten-Minute Thinking Tie-ins, Ten-Minute Real World Writing*

Entire contents copyright © 1997 by Monday Morning Books, Inc.
Box 1680, Palo Alto, CA 94302

For a complete catalog, please write to the address above, or visit our Web site:
www.mondaymorningbooks.com

e-mail address: MMBooks@aol.com

Monday Morning is a registered trademark of
Monday Morning Books, Inc.

Permission is hereby granted to reproduce
student materials in this book for non-commercial
individual or classroom use.

ISBN 1-57612-020-1

Printed in the United States of America
987654321

CONTENTS

Introduction   4

Activities

Animal Observation	6
Apparent Motion	8
Bouncing Balls	10
Constellations	12
Cycles	16
Decay	18
ESP	20
Evaporation	22
Fingerprints	24
Flotation	26
Gene Pool	28
Healing	30
Hearing	32
Heat	34
Lunar Craters	36
Memory	38
Mind Mapping	40
Motion	42
Musical Vibes	44
People Watching	46
Sight	48
Skeleton	50
Speech	52
Speed	54
Superstitions	56
Sweat	58
Taste	60
Teeth	62
Thinking Machine	64
Touch	66
Water	68
Waves	70

Skills

Classifying	72
Drawing	74
Experimenting	76
Fact Checking	78
Measuring	80
Model Making	82
Observing	84

Resources

Scientist's Guide	86
Ongoing Projects	91
Web Sites	94
Bibliography	95

Index   96

INTRODUCTION

Scientific discovery usually requires lengthy effort. For example, Gregor Mendel spent decades working out his theory of heredity. Robert Goddard invested a lifetime developing the science of rocketry. How then can students develop scientific skills in **ten-minute** activities?

The answer is simple, if not obvious. Science is a way of seeking truth through observation and experiment. While big breakthroughs are rare, the process is ongoing:
- When something out of the ordinary catches a scientist's eye, that's real science.
- When a scientist wonders about something, that's real science.
- When a scientist asks a question, that, too, is real science.

Ten-Minute Real World Science offers activities that last for only a few moments, yet evoke authentic scientific work. By engaging in many mini-projects, students gradually develop the skills of science.

Practicing Skills in Context

The key scientific skills are: observing, describing, questioning, experimenting, and reporting. These actions come into play whether a scientist is exploring gravity, animal behavior, sound, digestion, or any other topic. In other words, astronomers, biologists, chemists, physicists, sociologists, and all other scientists use the same basic approach to discover the nature of things.

Professional scientists usually specialize. But for the beginner, it makes sense to practice the skills in a variety of areas. This approach clarifies the nature of science and maximizes the chance that each individual will ultimately find an area of personal interest.

How to Use the Book

The activities in *Ten-Minute Real World Science* appear in alphabetical order. If you want to focus on a specific topic, for example, chemistry, use the Index.

Most lessons begin with a topic for students to ponder and then write about in their science journals. Background information for class discussion is provided in the margin.

The main activity is described in a few steps. Students are asked to observe and then, in their science journals, write or draw what they observed. In most cases, you'll need no materials or only items found in the classroom or at home. Occasionally, you'll find a reference to a Web site, such as one that offers magnificent lunar craters.

The **Skills** section offers activities to sharpen generic techniques, such as carrying out experiments.

Beyond Ten Minutes

An extension project accompanies each lesson. This material can be used to reinforce the given concept through independent study in school or at home.

Frequently, you'll find a list of library research projects designed to build students' general scientific knowledge and to relate their observations to a wide variety of issues—everything from cooking and bicycle riding to medicine and music.

The **Resources** section at the back of the book includes a reproducible Scientist's Guide offering tips on maintaining a science notebook, making sketches and so on. This section also contains a list of continuing projects, for example, a science book report series.

About Right Answers and Mistakes

Many science books tell students how to set up an experiment, and then immediately provide the right answer. This approach can be misleading, because in real scientific exploration answers are often elusive. Experiments may reveal nothing, or may have ambiguous results. Scientists spend lots of time in the dark. And they often reach false conclusions.

For example, when Galileo first observed the moon using a telescope, he was convinced that he saw vast seas. He also drew a large crater in the center of the lunar surface although no such crater exists. Galileo was wrong, but his method was right. By carefully reporting his observations, he set the stage for other scientists to correct his findings. The crucial thing was that he pursued new knowledge rather than relying on knowledge previously known.

This book attempts to simulate that situation so that young people can experience what it's really like to be a scientist. The aim is to encourage them to look closely at things for themselves, and to report their tentative observations, just as real scientists do.

Of course, students should be encouraged to check and question their observations. To help you guide them into self-evaluation, you'll find relevant, factual information in the margins.

Online Sharing

From time to time, we will supplement the materials in this book with free resources on our Web site: **www.mondaymorningbooks.com**. If you're not on the net, ask a colleague (or a student!) to download the items for you. If you have questions and comments, write us at **mmbooks@aol.com**.

ANIMAL OBSERVATION

Whales and other exotic creatures get lots of attention. But birds, ants, and other "ordinary" animals can provide fascinating opportunities for scientific study.

DIRECTIONS:
1. Remind students that scientific observation is purposeful. The scientist isn't just looking, but is looking for something.
2. Review the Animal Observation Guidelines, next page. Then have students study a classroom pet or one of the other candidates in the margin. During the session, students should take notes and draw pictures. (For drawing tips, see the Skills section.)
3. Afterwards, share the results. Discuss problems, ("My bird flew away!"), and brainstorm solutions. This could lead to reading books about animal watching. (See Bibliography.)

EXTENSION:
Have students conduct longer-term observations out of class and share results by posting drawings from their notebooks.

 A dramatic way to share observations is to imitate animal behavior, for example, walking like a dog or puckering the mouth like a fish. If students are interested in imitating bird calls, in addition to listening to bird live in nature, students might use recordings. The Internet is another source, for example, **www.birdsongs.com.**

Animals to Observe
ant
beetle
bird
cat
cockroach
dog
flea
fish
fly
frog
grasshopper
microbe
mosquito
moth
mouse
salamander
slug
spider
squirrel
worm

Animal Observation Guidelines

Scientific observation requires paying attention to details. In addition to describing things in words, you can often capture facts by drawing. For example, when studying an ant, try making a map of an ant's journey.

Think about the following points when you plan an animal observation report.

Observation information (include the following information in each observation report):
- name of observer
- type of animal observed
- name of species if known
- date of observation
- beginning and ending time of observation
- weather and temperature
- location of observation
- tools used, for example, magnifying glass

Physical description of animal
- color
- parts
- shape
- pattern of surfaces, for example, of a butterfly
- overall size (estimate if you can't measure)
- texture, for example, the shiny surface of a beetle

Locomotion
- method: crawling, flying, jumping, etc.
- speed
- path: straight, curved, zigzag, etc.

Actions observed
- building
- communicating
- digging
- drinking or eating
- fighting
- resting
- searching

APPARENT MOTION

Experiments with motion toys in the early nineteenth century led to many discoveries about vision. The same toys also paved the way to the invention of motion pictures and television.

MATERIALS:
- 12 note cards per group
- 1 pair of scissors per group
- 1 glue stick per group

DIRECTIONS:
1. Give students a copy of the Moon Phases Flip Book, next page.
2. Have them cut out the panels and glue them onto the bottom right hand corners of the cards. Let cards dry.
3. Students should then flip the pages of the flip book and observe the phenomenon.
4. Have students speculate on how the flip book works. As a hint, suggest that students think about electric signs that animate letters and pictures.
5. Optional: If you can get one or more strips of developed motion picture film, have students look at the sequence of frames.

EXTENSION:
Have students create their own flip books to illustrate science-related processes, such as a cell dividing, an inchworm inching, a solar eclipse, or an atom splitting. Advanced students might create computer animations using inexpensive software, such as *Dabbler 2* (Fractal Design). Related library research topics include:
- How are slow-motion movies made?
- How is an animated movie made?
- What is a stroboscope and how does it work?
- How do optical illusions work?

Movie Facts
- A movie projector shows 24 still pictures (frames) per second.
- The screen is briefly blacked out between each frame. For more than half the time, we're viewing a black screen.
- We "see" motion because each picture stays in our brain a short time and then fades while the next picture appears. The overlap creates the sense of motion.
- TV works in a similar way but uses 30 still pictures each second.

Moon Phases Flip Book

The moon seems to change shape each day of the month. These pictures show some of the changes.

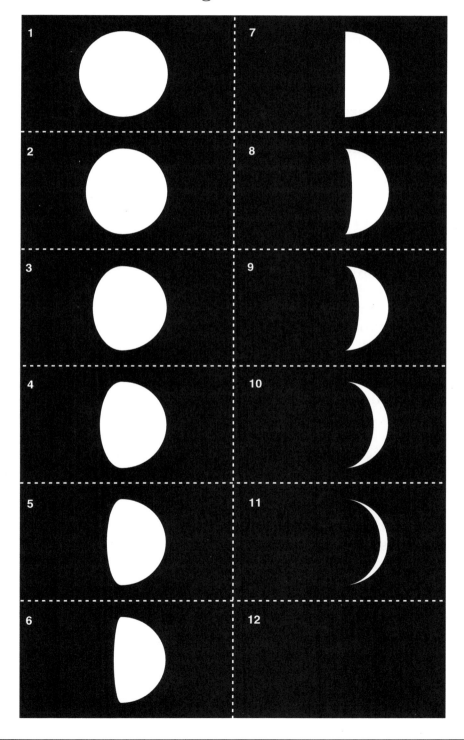

BOUNCING BALLS

Observing fast-moving objects requires concentration, persistence, and ingenuity.

MATERIALS:
- several identical tennis balls
- optional: other kinds of balls

DIRECTIONS:
1. Have students observe as you drop a tennis ball onto the floor. Then ask them to brainstorm questions about bouncing balls, for example:
- If a ball is dropped several times from the same height, will it always bounce the same number of times before coming to rest?
- Does the height from which a ball falls affect how many times it will bounce before coming to rest?
- Does each subsequent bounce come up shorter than the previous bounce? If yes, by how much?
- Do different kinds of balls bounce differently? For example, will a golf ball bounce as often as a tennis ball dropped from the same height?

2. Divide students into groups. Have each group choose a question, figure out how to answer it, and then try the experiment several times.
3. Each group should report its results.

Ready, set, drop!

EXTENSION:
Students can explore the rate at which falling bodies accelerate (gain speed) by using Galileo's method for studying motion, next page.

> **Notes on the motion experiment (next page):**
> The ball takes less time to cover the second half of the track. This shows that it moves faster as time passes. In order to observe this result, it may be necessary to adjust the angle of the board and the weight of the rolling object.

Galileo's Method for Studying Motion

When objects fall, do they gain speed? It's hard to tell because falling objects move quickly. But several hundred years ago, the brilliant scientist Galileo figured out a way to study this type of motion. He rolled a ball down a gentle slope. Because things fall slower that way, he could observe what was happening.

Try the following experiment to see if objects move faster as they fall. You will need a partner.

Materials:
- a plank
- a ball
- a watch that measures seconds
- something to rest the plank on
- a ruler

Method:
1. Set up the plank as in the picture. To keep the ball rolling straight, you might attach model train tracks or make your own tracks out of wood or metal strips.
2. Mark the midway point in the path.
3. Place the ball at the upper end of the track. Hold it there.
4. Have your partner keep track of time. When the second hand is at the zero point, have your partner say, "Go."
5. Release the ball, but don't push it.
6. When the ball reaches the midpoint, say "Midpoint." Your partner should call out the number of seconds that have passed.
Note: If the ball moved too fast to time, make the slope less steep.
7. When the ball reaches the end point, say, "End." Your partner should call out the total number of seconds.
8. Compare the number of seconds it took the ball to move through each half. What do you learn from the results?

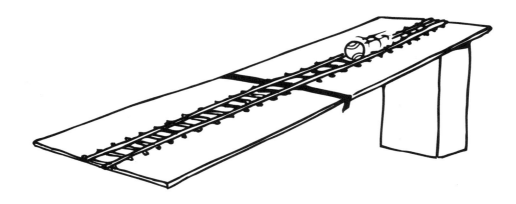

CONSTELLATIONS

Constellations are more than just entertaining pictures in the night sky. For thousands of years, these celestial patterns have served als an important astronomical tool. For example, Ursa Minor contains Polaris, the North Star.

DIRECTIONS:
1. Explain that constellations are star groups that, if viewed with a little imagination, form pictures of mythical people, animals, and objects. Astronomers and travelers have long used constellations to help locate specific stars.
2. Give each student or group a set of the Constellation Cards that have the constellations' names. Here, each constellation is presented with its key, or alpha, star.
3. After students have studied the labeled cards, give them the Constellation Cards without names to test their ability to recognize the constellation and its key star.

Star and Constellation Facts
- Under good viewing conditions (a clear night, away from city lights), a person with good vision can see about 2,000 stars with the unaided eye.
- There are 80 constellations in all.
- Some constellations are visible only from the northern or from the southern hemisphere.
- The time of year affects the visibility of many constellations.

EXTENSION:
Send the cards home so that students and parents can use them as star guides to find real stars. If you're online, you can find a remarkable Web site providing information, maps, and photos on all the constellations. Go to:
www.astro.wisc.edu/~dolan/constellations/

Constellation Cards

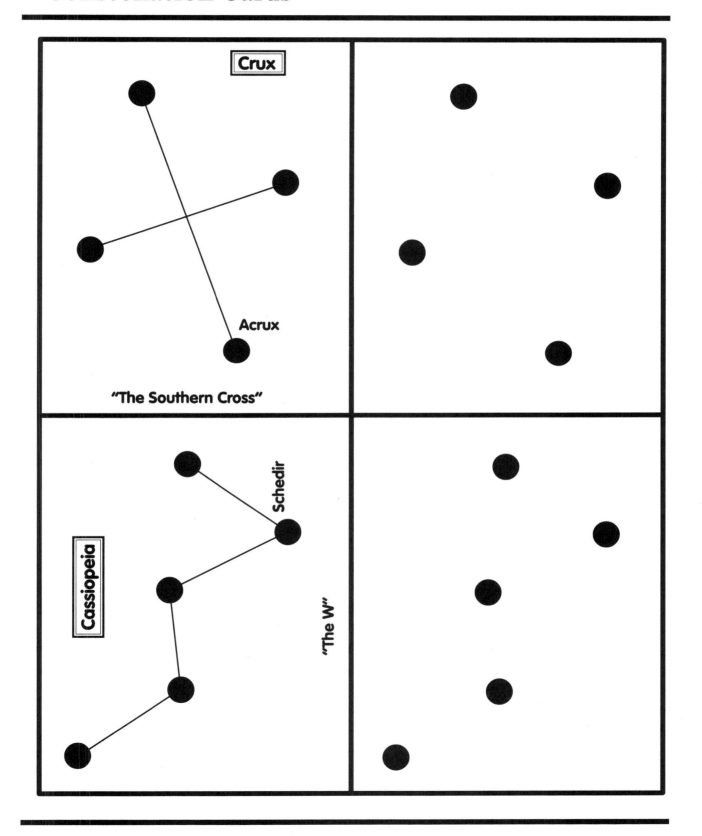

Constellation Cards

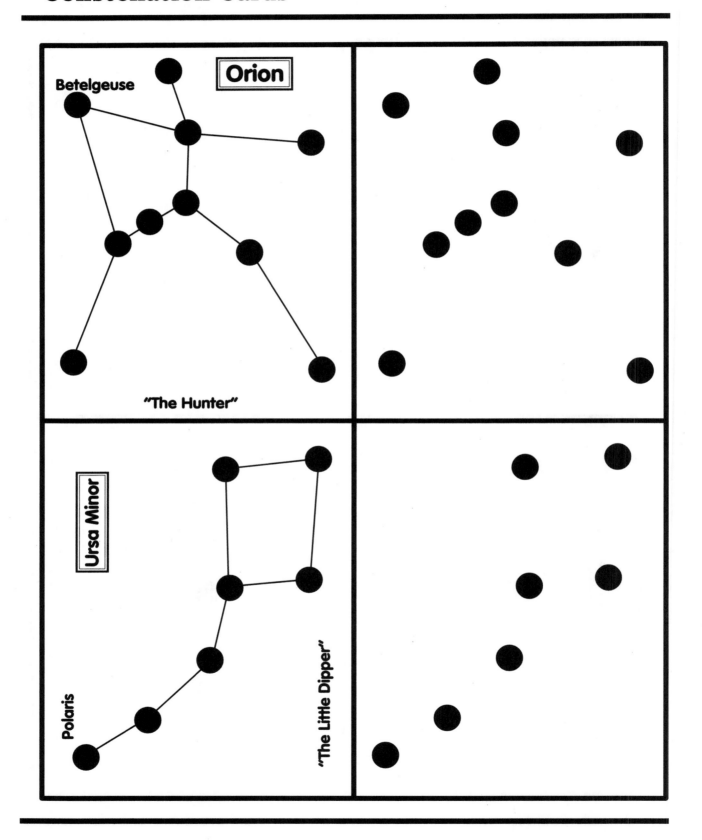

Constellation Cards

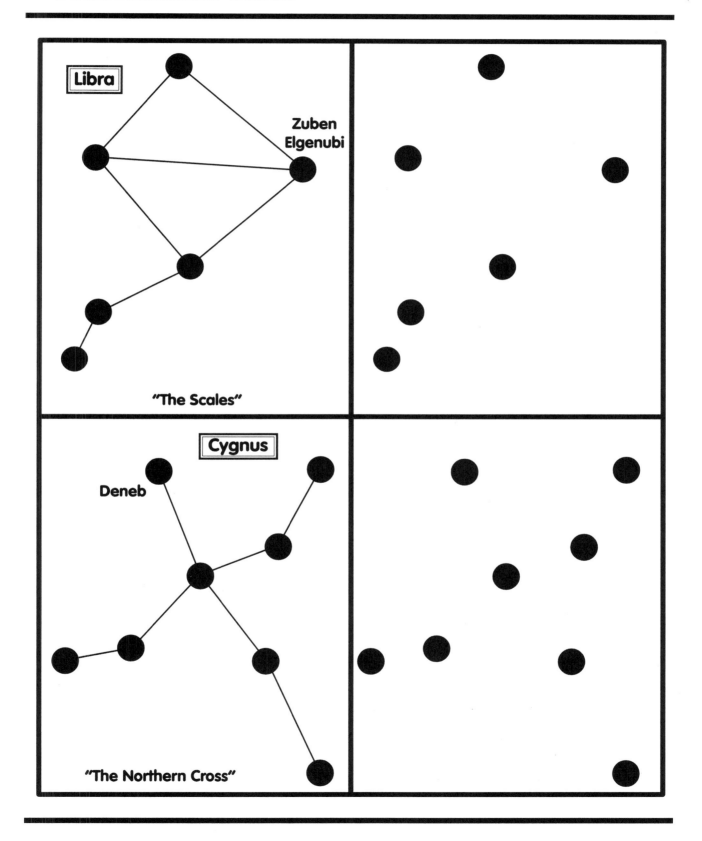

CYCLES

One of the first scientific achievements was the discovery of celestial cycles. This helped establish the idea that regularities in nature can be identified and often used.

DIRECTIONS:
1. Explain that a cycle is a repeating series of events. An example is day following night. Note that the word's root means "circle."
2. Tell students that they will study a cycle of great personal importance: breathing. For three minutes, they should write down everything they notice about it. They might try to answer such questions as:
• Where does the air flows in?
• Where does it flow out?
• In a quiet state, how many in-out cycles occur per minute?
• What parts of the body are involved?
(Hint: Students might rest one hand on the chest and the other on the belly.)
3. Discuss their observations. Explain that the diaphragm, a muscle located between the chest and the abdomen, expands the chest cavity and pulls in air. This muscle works without conscious control, though we consciously can hold our breaths for short periods, for example, when swimming under water.

Cycles to Study
bee colony
circulation of blood
comets
computer clock
eclipses
electric current
ice ages
insect life cycle
rain cycle
respiration (human)
respiration (plant)
seasons
skin cells
stock market
sunspots
temperature: daily
tides

EXTENSION:
Have students do library research on cycles, such as those listed in the column on the right. For a long-term, eyewitness project, students can use the Lunar Cycle Study activity, next page, to investigate the monthly waxing and waning of the moon.

Answers for next page:
• The lunar cycle takes about 29 days.
• The moon is visible during daylight hours for about 14 days in a row.

Lunar Cycle Study

The moon's shape seems to change day by day. At one part of the month, it's a thin crescent. The next day, the crescent is thicker. This process is called waxing. After a number of days, the moon looks round. Then it begins to grow thinner each day. This is called waning. Finally, the moon disappears altogether, and then appears again as a thin crescent.

This waxing and waning pattern is called the lunar cycle. To find out how long the cycle takes, try the following observation.

Method:
- Draw a grid with 40 boxes. Number the boxes.
- Look for the moon. Draw its shape in the first box. Write the date and time in the box. (Note: At some parts of the month, the moon is visible at night. Other times it will be visible in daylight.)
- Each day look for the moon and draw what you see. If you miss a day because of clouds or for other reasons, write the reason in the box instead of drawing the moon.

Questions to answer from observation:
1. How many days does it take for the moon to return to its starting shape?
2. Is there a pattern to when the moon appears in daylight? For example, does it happen every other day or for a certain number of days in a row?

Further study:
- Observe the moon at one time, for example, six o'clock in the evening. Then observe it an hour later. What change happened during that time?

DECAY

Rot is often considered yucky or bad. In reality, it is a vital phenomenon, and one that offers young scientists fascinating learning opportunities.

DIRECTIONS:
1. Ask students to write what they know about decay and rotting. Possible topics are:
• what decays and what doesn't
• how we recognize decay (smell, etc.)
• causes of decay
2. Review the facts of decay.
3. Ask students to observe as you place several organic and inorganic items in a covered box: a potato slice, an orange slice, a whole banana, a cup of milk, a piece of soggy bread, a plastic spoon. Have students predict in their journals what will happen to each item.
4. Have students inspect the items daily and describe, in words and pictures, what they observe.
5. Optional: Ask students to write stories that dramatize what would happen if nothing rotted.

Facts of Decay
• Everything that lives decays eventually.
• Some things decay faster than others.
• Decay is caused by tiny living things called bacteria and molds.
• As bacteria or molds eat dead material, they release gases that are the smells of decay.
• If nothing decayed after dying, once-living things would pile up.
• When things decay (decompose), their materials return to nature for use by living things. Thus, fallen leaves enrich the soil.

EXTENSION:
Try the Two Rotten Experiments on the next page. Related library research topics are:
• What keeps soil fertile?
• What happens to your town's garbage?
• What are bacteria and molds?
• How do bacteria cause illness?
• What is pasteurization?
• What bacteria live on and inside people?
• How can we avoid cavities and gum disease?
• How do freezing and canning preserve food?

Two Rotten Experiments

Burying Garbage

What happens to buried garbage? See for yourself.

1. Fill ten identical cups or buckets with equal amounts of soil. In each container, bury the same set of items. Include at least two items from each group.

Group A		Group B	
apple	grape	aluminum	nail
bread	newspaper	foil	plastic item
eggshell	wood	coin	stone

2. Divide the cups into two sets. Keep the soil in one set slightly damp. Allow the soil in the other to dry out.
3. Every two weeks, choose one cup from the damp set of cups and one from the dry set. Pour out the soil onto waxed paper and observe the items. Describe them in detail and compare those from the wet soil with those from the dry soil.
4. If you have a magnifying glass, look for mold growing on the garbage. If you have a microscope, look for bacteria.
5. Clean up carefully, and wash your hands when done.

Growing Mildew

Mildew is mold that devours things like cloth, paper, and bread. It grows best in dark, damp places. Here's a method for collecting and studying this mold:

1. Cut a piece of white cotton cloth into two pieces.
2. Soak each piece in water.
3. Set the pieces in the open air for a few hours. This will increase the chances of "catching" mildew spores floating in the air. Spores, like seeds, grow into new plants.
4. Seal each piece of cloth in its own plastic bag.
5. Store one bag in the dark, the other in the light. This way you can see which environment is best for growing mildew.
6. After three days, open each bag and observe it to see if any mildew has grown. Observing should include smelling. Describe your observations in detail, comparing the item stored in the dark with the one stored in the light.
7. Reseal the bags and check on following days.
8. If you want to see mildew clearly, you'll need to view it under a magnifying glass or, better, a microscope.
9. Clean up carefully and wash your hands when done.

ESP

Most scientists don't believe that extrasensory perception (ESP) has been proven to exist. However, many continue to study the topic scientifically.

DIRECTIONS:
1. Discuss the definition of extrasensory perception: communicating without using ordinary sensory contact. An example would be sending a thought directly from one person's mind to another's. Students might briefly hypothesize how such transmission might occur.
2. Use the ESP Test Cards, next page, to conduct a simple experiment. There are two methods.
• Method A: Choose a card at random, and study it for a set time (five or ten seconds) while students try to receive the image and draw it. Then go on to the next card.
• Method B: Divide the class into groups. One person acts as sender while the others act as receivers.
3. Discuss the results. If everyone involved knows what the cards are, by luck a receiver should correctly "get" the card 10% of the time. Of course, by luck someone might get many cards right during one trial. So a high score in one trial does not prove that ESP exists. That's why serious researchers insist on repeating this kind of experiment many times.

EXTENSION
In class or at home, students can try a classic Remote Viewing experiment. The procedure is:
• A person playing the role of "beacon" (sender) goes to a location unknown to a person or persons playing the role of "receiver." The beacon might be a teacher's aid, the principal, or an older student.
• At a set time, the beacon focuses on and draws something at the location for five minutes.
• At the receiving location, each receiver draws what comes to mind.
• The beacon's drawing is then compared with each receiver's drawing. In some cases, an independent judge scores the drawings for similarity.

ESP Test Cards

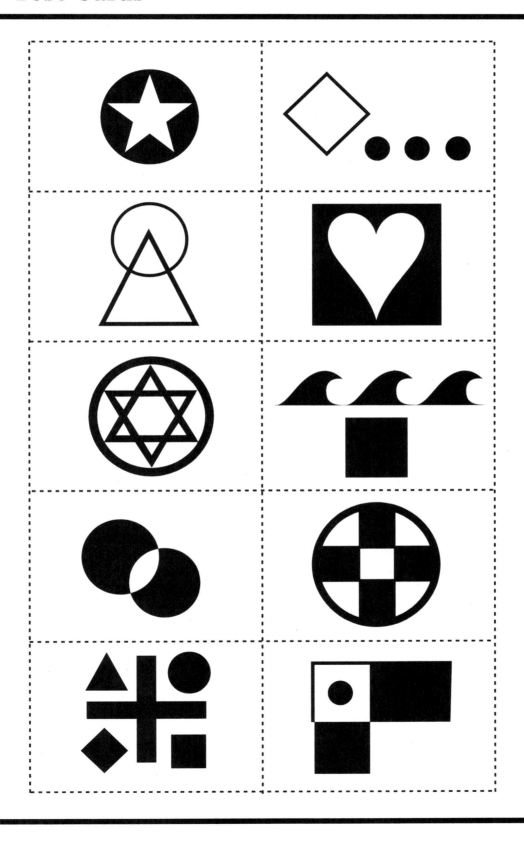

EVAPORATION

The change of water from liquid to vapor is a step in the important rain cycle.

MATERIALS:
- transparent plastic cups
- supply of clear water
- food coloring
- salt
- foil or plastic wrap

Evaporation Basics
- Water has three states: ice, liquid, and water vapor (an invisible gas).
- Evaporation occurs when liquid water turns to water vapor.
- The evaporation rate depends on four factors: water temperature, air temperature, amount of water vapor in the air (relative humidity), and wind.

DIRECTIONS:
1. If evaporation is a new subject for your students, review the facts on the right.
2. Ask students to predict the answers to the following questions, and then devise experiments to answer the questions:
A. When salt water evaporates, does the salt remain behind or disappear into the air?
   Experiment: Add a spoonful of salt to a cup of pure water. Place in a warm location. Observe what happens after the water evaporates.
B. Does salt water evaporate at the same rate as fresh water?
   Experiment: Pour pure water into one cup and an equal amount of salt water into another cup. Label the cups. Place in a warm location. Observe the rates of evaporation.
C. When water dyed with food coloring evaporates, does the dye stay behind?
   Experiment: Pour equal amounts of dyed water into two cups. Cover one cup. As water evaporates from the open cup, see if its color becomes darker than the color of the covered water. If it does, that indicates the dye doesn't evaporate.
3. Each day have students observe the cups and take notes.

EXTENSION:
Have students build and use a Solar Still as diagramed in the handout, next page. Other topics to explore through library research are:
- the rain cycle
- the formation of crystals
- the formation of stalactites and stalagmites

Ten-Minute Real World Science ©1997 Monday Morning Books, Inc.

Solar Still

A solar still is a device that uses sunlight to evaporate water from a liquid or from another material such as earth or vegetation. The device can provide drinkable water in an emergency.

Build your own solar still by following these directions. Then check the still daily to see how much water it produced.

Materials:
- rocks or other small weights
- water collector (cup, open can)
- leaves, grass, or other vegetation
- thin transparent plastic (refrigerator wrap)

Steps:
1. Dig a hole in the ground. The hole should be about a foot deep (.3 meters) and slightly smaller across than the piece of plastic you have available.
2. Place the water collector in the middle of the hole, open side up.
3. Place the vegetation around the collector.
4. Use the rocks or other weights to fix the plastic sheet over the hole. To make the set-up more secure, use small pegs to hold the plastic down. This sheet will allow the sun's rays into the hole to warm the air for evaporating the water. At the same time the plastic will keep the warm, moist air from escaping.
5. Place a small weight in the center of the plastic, so that the plastic sheet drops slightly in the center above the water collector. As the water evaporates from the vegetation and from the earth, it will rise to the plastic sheet, then condense, and roll down the plastic to the lowest point, where it will drip into the collector.

FINGERPRINTS

The use of fingerprints in crime fighting illustrates the value of careful observation. The following activity puts the focus on noticing distinctive details.

DIRECTIONS:
1. Give each student a copy of the Fingerprint Matching handout, next page.
2. Discuss the theory that each person's fingerprint pattern is unique: that is, it differs from those of every other person in the world. Ask students how this hypothesis might have developed? What would it take to disprove it?
3. Challenge students to match each numbered print with a corresponding lettered print.
4. Have volunteers draw pictures on the board indicating which features they used to make the matches.
5. Optional: Speculate on why fingerprints exist. As a hint, ask students to think of other things with rough surfaces, for example, a pair of pliers with grooved jaws. How could the students' hypotheses be tested? (Try doing fine motor tasks, like threading a needle or catching a ball, while wearing rubber gloves.)

EXTENSION:
Related library research projects are:
• the use of DNA for identifying individuals
• the system of classifying fingerprints in a database so that prints can be traced back to individuals

Answers for next page:
1 matches B. 2 matches A. 3 matches C.

Fingerprint Matching

Fingerprints are made up of ridges in the skin. No two people have exactly the same fingerprints. That's why it is possible to identify people by their fingerprints.

The six thumbprints shown here came from three people. Each numbered print matches one lettered print. Carefully observe the prints. Then try to match a numbered print with a lettered print.

Note: Each person was fingerprinted twice. For this reason, when you find a match between a number and a letter, you may see a slight difference in darkness or position. But each person's basic fingerprint pattern will be the same.

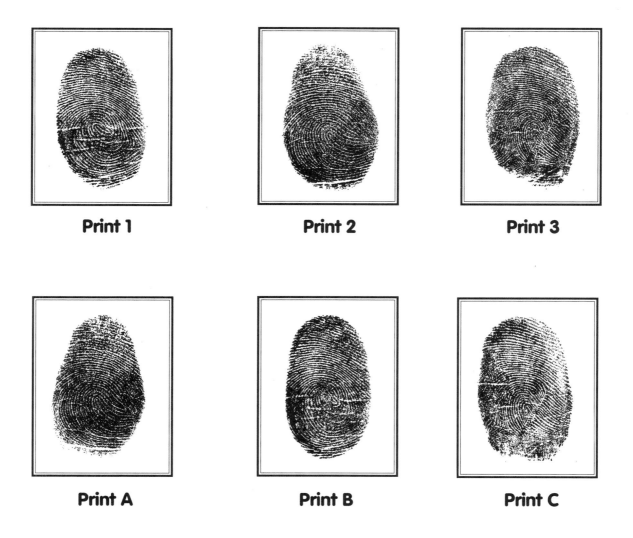

Print 1 Print 2 Print 3

Print A Print B Print C

FLOTATION

One of history's great moments occurred when the ancient Greek scientist Archimedes discovered why things float. His shout, "Eureka" (I found it), has echoed through the ages.

MATERIALS:
- a large glass baking dish
- a marble-size glob of clay for each student

DIRECTIONS:
1. Have students hypothesize about why some things float and others sink. Encourage them to use drawings to explain their hypotheses.
2. Review the basics of flotation.
3. Drop the glob of clay into the water to demonstrate that it sinks.
4. Give each student a piece of the clay.
5. Ask students to figure out a way to shape the clay so that it will float. (Boat shapes work best!)
6. Have students test their constructions.
7. Optional: Have students add weight to their boats to the point of sinking them.
8. Discuss the results.

EXTENSION:
Have students conduct other flotation observations, such as those outlined on the next page. Topics for library research include:
- Why are icebergs so dangerous?
- What makes hot-air balloons fly?
- What happens to smoke from a chimney?
- Why do swimmers float?
- Why does an overloaded boat sink?
- What makes a submarine sink and then return to the surface?

Flotation Basics
- An object in a liquid will be pulled down by gravity.
- As it sinks, the object pushes liquid aside. This is called "displacement."
- The displaced liquid pushes back.
- An object will float when its weight equals or is less than the weight of the water it displaces. If it floats, its density (mass divided by volume) is less than the liquid's density.
- For example, a liter of lead weighs more than a liter of water, and will sink, whereas a liter of cork weighs less than a liter of water and floats.

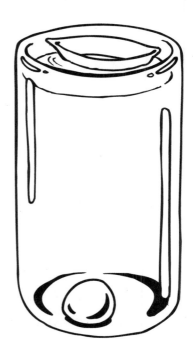

Flotation Experiments

Air
• Fill a glass halfway with water. Place a drinking straw in the glass so that one end rests near the bottom of the glass. Gently blow into the straw. Describe what happens to the bubbles. Why do you think it happens?
• Fill a jar nearly to the top with water, leaving a small amount of space at the top. Seal the jar tightly. Turn the jar over. What happens? Why do you think it happens?

Ice
• Place an ice cube in a glass of water. How much of the ice is above the surface of the water?
• Place the same ice cube in a glass of cooking oil. How much of the ice is above the surface of the oil?
• Place the same ice cube in a glass of molasses. How much of the ice is above the surface of the molasses?
Explain the results of these experiments. Can you guess which weighs the most: an equal volume of water, of oil, or molasses?

Oil and Water
• Pour equal amounts of water and oil into a glass. What happens?
• Cover the glass and shake. What happens right away? What happens after a few minutes?
• Describe a method that might be used to remove oil from a lake or ocean if it spilled from a ship.

Molasses
Pour an inch (2.5 centimeters) of molasses in a glass. Add the same amount of water and the same amount of cooking oil.
• What happens? Which liquid is on top, which in the middle, which at the bottom?
• Gently lower a plastic game piece (checker, small building block) into the glass. Where does it end up? Why do you think that happens?

GENE POOL

Genetics will be among the major areas of science in the next hundred years. The following activity brings the subject close to home.

DIRECTIONS:
1. Discuss the meaning of "heredity."
2. Draw a simple branch diagram, such as the one below that illustrates how family members contribute to a child's genetic makeup.
3. Ask students to use the branching diagram to figure out how many people have contributed to their genetic background starting with their great, great, great, great-grandparents.
4. Optional: If you're studying exponents, have students calculate how many ancestors contributed to their gene pools going back 10, 20, or more generations. The first generation would be 2^1 (2 parents); the second would be 2^2 (4 grandparents), and so on.

EXTENSION:
Have students construct family circles such as the one on the next page.

> **Heredity Basics**
> • Heredity refers to the transfer of biological information that helps shape each individual.
> • The information is contained in chemical structures called genes.
> • Humans have approximately ten million genes.
> • Each person's ancestors contributed to that person's genetic background, known as the person's gene pool.

Ten-Minute Real World Science ©1997 Monday Morning Books, Inc.

Family Circles

The family circle diagram displays the ancestors whose genes helped shape you. Make your own family circle using the following steps and model.

Step 1. Draw a small circle. Write your name in it.

Step 2. Draw a second circle around the first circle. In the left half, write your mother's name. In the right half, write your father's name.

Step 3. Draw a third circle around the first two circles. Divide the left half of this circle into two parts. In one, write the name of your mother's mother (your maternal grandmother). In the other left-hand part, write the name of your mother's father (your maternal grandfather). If you don't know their names, ask a relative.

Step 4. Do the same thing on the right half of the circle, using the names of your father's parents.

Step 5. Add another circle. Divide it into eight parts, one for each of your great-grandparents. Fill in their names.

Step 6. Continue this way until you run out of room or out of information.

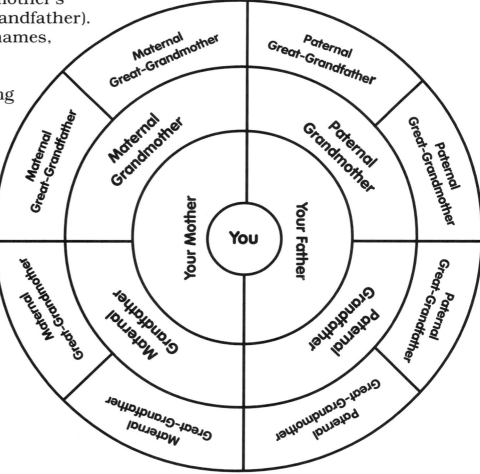

HEALING

We often take the body's self-healing abilities for granted. Yet this phenomenon is one of the nature's most remarkable happenings. The following activity aims to help children understand that things like scabs and pus are remarkable, not yucky, and are worthy of thoughtful study.

DIRECTIONS:
1. Have students think about their own experiences with skin cuts, and then use scientific (non-emotional) words to define *pus*, *scab*, and *scar*. Students might also briefly list the steps in the healing of cuts.
2. Review the definitions, using the information in the margin.
3. Give students a copy of the Steps of Healing play, next page.
4. Divide the class into small groups.
5. Have students read aloud the play in the small groups.

EXTENSION:
Encourage students to observe the healing of minor skin injuries that happen to them or family members. They should document the healing day-by-day in their notebooks. Related topics for library research are:
• first aid for minor skin injuries
• how antiseptics work
• how the body responds to viruses
• how starfish regenerate lost limbs
• the value of the body's regurgitation reflex

Healing Words
• *infection*: a disease resulting from the presence of viruses or microscopic plants and animals
• *pus*: the yellowish liquid matter produced by infection, and made largely of white blood cells and bacteria
• *scab*: a crust over a wound during healing, made of blood cells
• *scar*: a mark left on the skin after a wound
• *white blood cell*: a colorless cell that makes up part of the blood and that battles infections by killing bacteria

Steps of Healing

Person: Ouch. What happened?

Skin: I got cut when you fell off your bike and hit that rock.

Person: Yikes. There's blood all over.

Blood: No wonder. The rock not only tore the skin. It also broke the tiny tubes called blood vessels under the skin. These tubes carry blood to skin and nerve cells.

Person: Blood is gross.

Blood: I am not. I'm just doing my job. By bleeding I wash the dirt out of the cut.

Person: I think you've washed it out enough. Can't you stop?

Blood: Don't worry. When blood cells contact air, they form a clot. This stops the bleeding. The clot becomes a scab.

Person: Scabs are yucky.

Scab: Think again. I protect the wound from germs. Without me, you might get an infection.

Blood: If an infection does happen, white blood cells gather and attack the germs. It's like a war.

White cell: The bodies of white cells and germs form pus.

Person: That's really gross.

Pus: No, it's natural.

Skin: Meanwhile, I'll be growing new cells that fill in the cut. These cells eventually make the skin surface whole again. Don't tell me that this is gross.

Person: I won't. Now, I see that it's amazing!

HEARING

Investigating the sense of hearing can lead to a wide variety of topics, ranging from high fidelity to sonar to self-awareness.

DIRECTIONS:
1. Ask students to briefly describe the process by which they hear the words of someone talking to them.
2. Tell students that they're going to do an experiment concerning the sound of their own voices.
3. Show them how to cup their hands behind their ears with fingers together.
4. Divide the class into groups. Have each student take a turn saying a few words normally ("Testing, testing"), and then repeating the same words with the hands cupped behind the ears.
5. Students should describe the results and try to explain any differences between the two methods. See the explanation in the margin.
6. Optional: If you have a tape recorder, have students tape their voices. Voices taped from the air should sound like the voices heard with the cupped hands.

EXTENSION:
To test the ear's ability to distinguish sounds, record a variety of sounds, such as: zipping a zipper, tearing paper, stirring a liquid, snapping your fingers, sharpening a pencil, putting a key in a lock, rubbing sandpaper on wood, cracking an egg. Play the tape and ask students to identify each sound.

Hearing Via Air and Bone
- The sensation of sound is caused when energy vibrates our hearing receptors inside our skulls.
- When we hear our own voices, much of the sound reaches our hearing receptors through the bones in our skulls.
- Other people's voices come through the air.
- By cupping our hands, we collect sound waves from the air and hear ourselves as others hear us.

Hearing Experiments

Hearing Cone
Most sounds that we hear travel as waves through the air. Our outer ears funnel some of the passing waves into the inner ear, where they are turned into electrical signals. These signals then go to the brain. If our outer ears were as big as those of a rabbit, would we hear better? Find out through an experiment.
1. Roll a piece of drawing paper or newsprint into a cone.
2. Hold the small end of the cone near the ear. But do NOT put the end of the cone or anything else into the opening of the ear.
3. Compare listening with and without the cone.

Decibel Hunt
Loudness is measured in decibels. You can learn how to recognize decibel levels by listening for the following sounds.
- 5 decibels: pin drop
- 50 decibels: normal talking
- 20 decibels: ticking watch
- 90 decibels: urban traffic

■ ■

Sounds of 100 decibels can damage hearing. That's why people who work near loud sounds wear ear protection. To protect your hearing, avoid listening to loud music or sounds.

Sound Finding
Can people tell where sounds are coming from? Find out with a simple experiment. You'll need a partner.
1. Blindfold the subject. Tell the subject to try to point in the direction of a short sound (hand clap) or your voice.
2. Quietly move about a meter (3 feet) from the subject.
3. Clap your hands once. After the subject points, say a few words and see where the subject points.
4. Quietly move to another position and repeat the test.

HEAT

Thermodynamics is the science of heat and related forms of energy. The word is big, but many of the issues can be studied at a simple level.

DIRECTIONS:
1. Have students list words related to "heat," for example: sun, fever, fire, steam, cold, and thermometer.
2. To stimulate more ideas, conduct a simple experiment: Have students rub their hands vigorously, generating heat through friction, and then describe the results in their journals: what happened and why?
3. Optional: Fill a glass halfway with hot water (not boiling), and stand three items in it: a metal knife, a plastic knife, a wood pencil. Have students touch the top of each item, then describe and analyze the results.
4. Using the above experiences, have students write definitions of "heat." Encourage them to go beyond a circular definition, such as "Heat is something hot."
5. Discuss the students' ideas, and add information from the Heat Basics box.

EXTENSION:
After trying the insulation project, next page, students might do library research on heat-related topics such as:
• Where does the sun's heat come from?
• What causes a fever?
• How does aspirin work to lower fever?
• Do all mammals have the same normal temperature that humans have?
• Does hibernation change an animal's temperature?
• What is cryogenics?
• How does natural gas produce heat?
• Compare the Fahrenheit and Celsius scales.

> **Heat Basics**
> • Heat is a form of energy: the ability to change things.
> • The faster a molecule vibrates, the more heat it has. (A molecule is a bit of matter.)
> • Other energy forms can change into heat. For example, when a light ray hits an object, some energy becomes heat. We feel hotter in the sun than in the shade. Sun rays cause skin molecules to vibrate faster.
> • Conduction occurs when heat moves through objects. If the bottom of a metal spoon sits in hot water, the top will eventually feel hot.
> • Some objects conduct more heat than others. Poor conductors are called insulators.

They're getting warmer!

Build an Insulated Container

We often try to control the flow of heat. This is what you do, for example, when you wear a heavy coat. The coat insulates you from the cold by slowing the flow of heat from your body. By building an insulated container that keeps liquids hot or cold, you can better understand how heat-control works.

Materials:
- large milk carton (.5 gallon/2 liters)
- two small cartons (1 pint/half liter)
- piece of cork or wood
- glue or tape
- ice water
- large pan

Procedure
1. Completely open the tops of all three cartons.
2. Tape or glue the cork or wood to the bottom of one small carton to make a base. Cork or wood slows the flow of heat.
3. Place the carton with the base into the large carton so that it stands in the middle of the large carton. **This carton-within-a-carton is your insulated container.**
4. To test your insulated carton, fill a pitcher with warm water. Pour equal amounts of the water into the two small cartons.
5. Fill the pan with ice water.
6. Place the insulated container and the other small container into the pan with the ice water.
7. After a few minutes, feel the water in each small carton. Which is warmer? If you have a thermometer, you can more accurately check the success of your insulated container.
8. Try to figure out ways to improve the insulated container.

LUNAR CRATERS

Long before astronauts reached the moon, scientists tried to figure out how the craters were made. One early research method was to use simulation, a technique of great value for modern scientists.

MATERIALS:
- containers filled with sand or flour
- flour
- water
- marbles

DIRECTION:
1. Divide the class into small groups.
2. Give each student a copy of the Lunar Craters handout, next page.
3. Have students speculate on how the craters might have been formed. Or ask them to decide between the two major theories:
- craters resulted from volcanoes
- craters resulted from meteorite crashes
4. Explain that a simulation is an experiment that imitates nature. In the seventeenth century, Robert Hooke used a simulation to determine if craters were made by meteorites crashing into the moon.
5. Have students repeat Hooke's simulation:
- Create a simulated lunar surface from flour or sand.
- Make a flour-ball meteorite model, or use a marble.
- Drop the "meteorites" onto the "lunar surface."
- Compare the impact area to pictures of real craters.
- Decide if this kind of impact could produce craters.
6. Discuss the results of the experiment. Today, many lunar scientists believe that some craters were formed by impact, and others by volcanic eruptions.

EXTENSION:
Arrange for students to study the moon's major craters using binoculars or a telescope with a power of 50 magnification or more. Possible sources of equipment will be a local amateur astronomer or a high school or college astronomy club.

If you're online, you can find many Web sites featuring dazzling lunar crater photos. A good example is sponsored by Grove Greek Observatory, in Australia. You'll find them at
http://www.netsys.com/~steven/pictor.html

Lunar Craters

The word crater comes from an ancient Greek word meaning "bowl." Craters often are shaped like bowls.

Thousands and thousands of craters cover the moon. These range in size from a few inches (centimeters) to more than 100 miles (160 kilometers) across.

Some craters have flat center areas. Others contain mountain peaks in the middle.

Bright lines radiate like spokes from some large craters. These lines are called rays. In a few cases, the rays are more than 1,000 miles long (1,600 kilometers). The rays are best seen when the moon is full.

In addition to the moon, some planets are covered with craters. There are even a few moon-like craters found on Earth.

MEMORY

Memory plays a key role in education and personality.

DIRECTIONS:
1. Ask students to describe memory as if explaining it to someone who never heard of it. They might how memory is used where it occurs. (Some ancients thought that the heart was the organ of memory! Hence the phrase "Learn it by heart.")
2. Ask students to predict in their journals the longest number (in digits) they can recall if they hear the number one time. (Memory scientists have found that 7 or 8 digits is the limit for most people.)
3. Have students label rows in their journals, A to P.
4. Slowly read the numbers on the next page. Students should listen to each number but not write it until you've read all the digits of the number.
5. After reading a number, give students two or three seconds to write it before reading the next number.
6. When done, read the list of numbers again and have students check those they got right.
7. If time permits, graph the number of right and wrong answers, and see if there is a pattern. Optional: Have students repeat the experiment but this time with visual stimuli using the cards on the next page. They should stack the cards number down, then turn over a card, read it for two or three seconds, put it aside, and write the number.

Memory Basics
- We have short term and long term memory.
- Short-term memory lets us repeat a name we just heard. But if we don't focus on the name, we may forget.
- Long-term memory stores facts, habits, and everything else that we learn. Some scientists say that these memories stay with us even if we can't always recall them. Others believe that long-term memories can change.

EXTENSION:
Encourage students to devise and carry out memory experiments to answer questions, such as:
- When trying to memorize facts (such as capital cities), is it better to play music or to have silence?
- When trying to memorize facts, does it help to go to sleep right after studying (to sleep on the information)?

Memory Test Cards

A. 638	B. 19	C. 8940	D. 27931
E. 96184	F. 78243	G. 58676	H. 217408
I. 625397718	J. 6741	K. 2054132	L. 54861893
M. 9460031	N. 13806	O. 38585207	P. 71429745

MIND MAPPING

Exactly how the mind works is still a puzzle. But most scientists believe that the linking of ideas is a key part of the process, especially in terms of memory and creativity.

DIRECTIONS:
1. Explain that one way scientists explore thinking is by word association. A scientist gives a word to the subject (the person whose thinking is being observed). The subject then responds with the first word or phrase that comes to mind.
2. Tell students to write down the first word or phrase that comes to mind when you say a word. Allow about five seconds between words. Examples of words to use appear in the margin.
3. Give students a chance to share their associations. But first establish two rules:
• No one has to share.
• No associations should be criticized.
4. Variation: Give a single word, and see how many associations students (alone or as a group) can make to it. For example, the word "oil" might link to cooking, well, painting, spill, and bicycle chain.

Words to Associate
air
bicycle
book
clock
food
friend
park
pet
plastic
pollution
sports
television
tree
window

EXTENSION:
Have students use webbing to map associations in greater depth. See the model, next page. Point out that this kind of mapping is useful in developing ideas for a report or story. Related library research topics are:
• amnesia
• memory tricks (mnemonics)

Ten-Minute Real World Science ©1997 Monday Morning Books, Inc.

Sample Mind Map

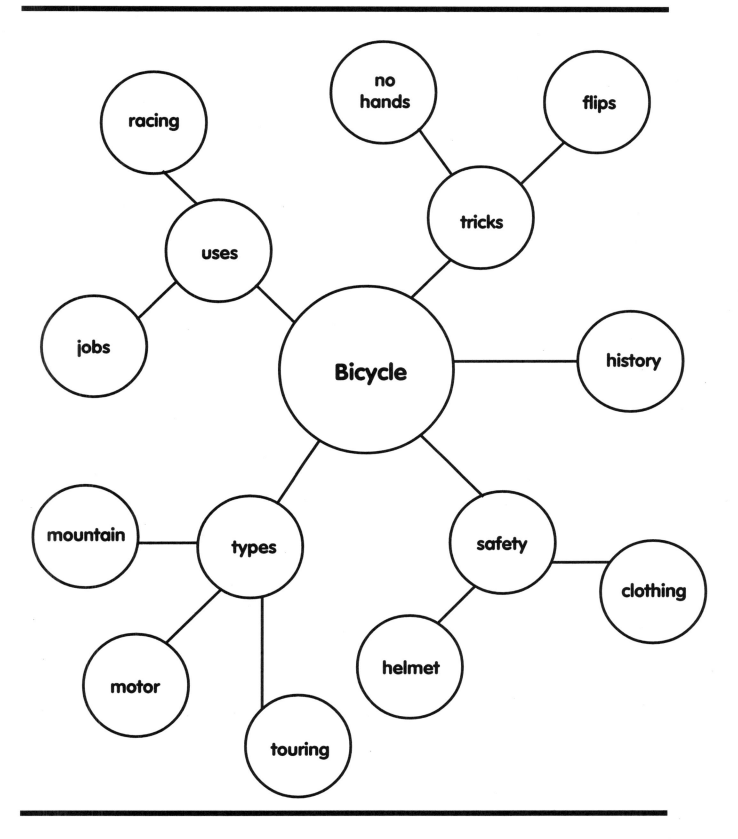

MOTION

Newton, Einstein, and other physicists devoted much effort to studying motion. Your students can join in this work by observing everyday activities, such as bicycling and swimming.

MATERIALS:
- thin disks (coins, metal washers, bottle caps)
- optional: toy tops

DIRECTIONS:
1. Divide the class into groups. Give each group a thin disk.
2. Have students try to balance the disk on edge while it's stationary and again while it's rolling. If you have toy tops available, students should try to balance them when the tops are rotating and when not rotating.
3. After recording their observations, students should try to explain the results.
4. Discuss the results, using the information in the box on the right.

EXTENSION:
Have students read Newton's Three Laws of Motion (next page) and answer the questions. Related library research topics include:
- What makes rockets move?
- Why does a roller coaster climb the second hill?
- How do worms move?
- Why does the moon stay up?
- What makes food move through the body?

Balance Basics
- In most cases, students will find that the disk is easier to balance while rolling. Similarly, a top is easier to balance while it's spinning.
- These results relate to observations made by Isaac Newton in the seventeenth century. Newton wrote that things in motion tend to continue moving in the same direction. The rolling coin's forward motion contributes to its stability.
- In the same way, a bicycle is easier to balance when it's moving than when it's at rest.

Answers for next page:
1. Law 3   2. Law 1.   3. Law 2.

Newton's Three Laws of Motion

In the seventeenth century, the great English physicist Isaac Newton discovered three laws that describe how things move. Read the laws, then answer the questions below.

Law 1. If something is moving, it will continue to move in the same direction unless something stops it or changes its direction. If something is motionless, it will stay motionless unless something moves it.

Law 2. The force an object has is related to the object's mass (weight) and its acceleration (change in speed).

Law 3. If you push in one direction, there is an equal and opposite push in the other direction.

Questions:
1. Which law explains why a ball bounces?
2. Which law explains why it's hard to stop running when you're running down a hill?
3. Which law explains why dropping a bowling ball on your foot hurts more than dropping a golf ball on your foot?

MUSICAL VIBES

Music is an art. But it is also a subject that has long fascinated scientists.

MATERIALS:
• a shoe box top or a rectangular piece of cardboard or wood (for the banjo's "body")
• a rubber band big enough to fit around the body
• a piece of chalk, a pencil, or other small item (for the "bridge") to lift the rubber band so it can vibrate

DIRECTIONS:
1. With the class, brainstorm a list of musical instruments. Then ask students to write their ideas about how musical sounds are made. For example, what makes some notes sound higher (have a higher pitch) than others?
2. To demonstrate that music involves vibrations, have students hum high and low notes into their palms. What do they feel?
3. Explain that most musical instruments have parts that vibrate. For example, in a saxophone, a wood reed vibrates. In a piano, strings vibrate. In a human being, voice sounds are created when air vibrates muscular strips (vocal cords) in the throat.
4. Choose a simple and familiar melody, for example, "Row, Row, Row Your Boat."
5. Have students softly "play" the tune using their vocal chords, but without singing words. They should observe the changes that take place as the pitch changes.
6. Discuss the results.

EXTENSION:
Have students make rubber band banjos, and experiment with the pitch by pressing down to shorten (and tighten) the length of the band. Related library research topics are:
• How are sounds produced by the flute, the trombone, the organ and other instruments?
• What is a musical scale?
• How do birds sing?
• How does a telephone reproduce voice?
• Why do musical instruments sound different when playing the same melody?
• Why don't people's voices sound alike?

Pitch Basics
• All sounds, including musical sounds, are created by vibrations.
• The more vibrations per second, the higher the pitch of the sound.
• If a vibrating string is tightened, it will vibrate faster when plucked or bowed, and hence will have a higher pitch.
• Tones created by the human voice change pitch when the muscles in the voice box tighten or relax.

Rubber Band Banjo

The banjo belongs to the family of musical devices called stringed instruments. Members of this family include instruments whose strings are plucked (the harp, the guitar); instruments whose strings are vibrated with a bow (the violin, the cello), and instruments whose strings are tapped (the piano).

To make your own stringed instrument, study the diagram on this page. Then experiment with the instrument. Here are some ideas:
- Play different notes by pressing down against the rubber band as you pluck it.
- Compare the tones made when you pluck the rubber band while pressing it in the middle and when you pluck it without pressing it.
- Try to play a scale or a simple melody.
- Add a second rubber band with a different width. Does the width make a different kind of sound?

PEOPLE WATCHING

The study of people in faraway places made anthropology famous. But this science can be applied to human behavior anywhere.

DIRECTIONS:
1. Give students the People Watching Guide, next page. Discuss the goal of anthropology: to learn about everyday human behavior.
2. Choose a school location for a brief, people-watching experience. Possibilities include:
- the playground at recess
- the lunchroom
- the gymnasium
- a classroom of younger students (your anthropologists might visit it in teams of two)
3. Review the people-watching tips. You might do a practice run with one group of students acting as the subjects, perhaps doing a reading circle, and the rest of the class observing.
4. Conduct the observation.
5. Discuss the results.

EXTENSION:
Have students watch several episodes of a TV series and write an anthropological report on the lives of the characters as if they were real people. Topics might include:
- clothing
- manners
- pastimes (hobbies, games)
- tools used
- work

The report might compare the fictional world with the real world as students know it.

People Watching Guide

Scientists who study people's lives are called anthropologists. Their science is called anthropology. It covers topics such as work, dress, manners, games, food, health, and religion.

With most sciences, the subjects don't know that they are being studied. For example, a planet is not "aware" that an astronomer is looking at it. Anthropology is different. When people know that they are being observed, their behavior may change. For this reason, anthropologists must be careful about how they make their observations. Here are some suggestions.

1. Keep an open mind. Don't judge people if their behavior seems different from yours.

2. Stay in the background. If you're trying to observe people without disturbing them, try the following:
- Watch from a distance.
- Don't talk or make loud noises.
- Pay attention to details.
- Record notes (including pictures) in a small notebook.

3. Focus on behaviors. These might include:
- eating
- laughing
- talking
- crying
- looking at things
- making things
- playing
- reading
- shopping
- working

4. Look for details. These can include physical facts about dress and jewelry. It also includes information about actions. For example, when people talk, do they:
- Make eye contact?
- Use hand gestures?
- Stand close together or far apart?

5. Record information about objects that people carry. Examples are books, bags, and tools.

6. Notice dress, jewelry, and other physical details.

SIGHT

The sense of sight offer many research topics.

MATERIALS:
• large colorful object, such as a piece of wallpaper or an art print (hidden from view)

DIRECTIONS:
1. Ask students to write what they know about the sense of sight.
2. Divide the class into pairs. Give each team a copy of the Eye Diagram, next page. Have students use the diagram to identify the parts in the partner's eye. CAUTION: Tell students they should never touch their own or others' eyes!
3. As time permits, try these experiments:
   Light Level Experiment: Ask students to observe what happens to their partner's pupils when you darken the room and then, after a few minutes, lighten it. Students should enter their findings in their notebooks and speculate on why the pupils dilate (got bigger) and contract.
   Color Vision Experiment: Dim the lights. Bring out the colorful object. Ask students to describe it. Turn on the lights. Repeat the description. What hypotheses can they make about the results?
   Parallax Experiment: Have students stretch their arms at full length. They should hold up their thumbs and align them with an object across the room. They should close one eye and then the other. What happens?

EXTENSION:
At home, have students study the eye blink rate of family members. Related topics for library research include:
• What causes blindness?
• What causes color blindness?
• What are optical illusions?
• How is a camera like and different from the eye?

Sight Basics
• Light level: Light enters the eye through the *pupil*, a small hole at the front. Because there is no light inside the eyeball, the pupil appears dark for the same reason that a window in a darkened house seems dark when viewed from outside.

The *iris*, a colored circle around the pupil, contains muscles. In bright light, the muscles contract and shrink the pupil. In dim light, the muscles relax, enlarging the pupil to let in more light. As the pupil opens, we become used to the dark.

• Color vision: We have two types of cells that respond to light. Rods, which are sensitive to shape but not color, work in dim light. Cones, which are sensitive to color, work less well in dim light. That's why colors seem to fade in dim light.

• Parallax: Because our eyes are separated, they look at objects from a different angle. We don't sense this with both eyes open because the two images blend.

Eye Diagram

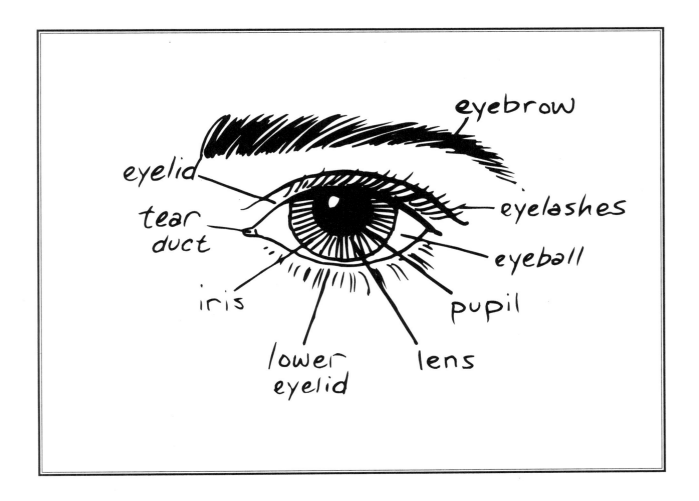

SKELETON

Skeletons are often a sign of death and decay. But these intricate structures deserve a far more positive appraisal.

DIRECTIONS:
1. Ahead of time, attach a sheet of white paper along a classroom wall.
2. First thing in the morning, have partners mark each other's height by drawing a line on the sheet, labeled with the measured student's name plus the time designation "a.m." Include yourself and other adults if you like.
3. Near the end of the day, repeat the process. Usually, the second measurement will indicate a decrease in height.
4. Have students discuss possible reasons for the results. See if anyone guesses the true fact that gravity pulls down on the body, squeezing the vertebrae together.

Scientific name	Everyday name
carpals	wrist bones
clavicle	collarbone
cranium	skull
femur	thighbone
mandible	lower jaw
patella	kneecap
pelvis	hipbone
radius	forearm bone
sternum	breastbone
tarsals	ankle bones
tibia	shinbones
vertebrae	spine/backbone
zygoma	cheekbone

EXTENSION:
Give students a copy of the Skeleton Diagram, next page. Have them try to identify the common names used for some of the bones. The answers appear in the box on the right. Related topics for library research include:
• How is the human skeleton similar to, and different from, a skyscraper's frame, or the frame of an automobile?
• How is the human skeleton similar to, and different from, the skeleton of a dog, a bat, or another animal?
• What animals have exoskeletons (insects, crabs, lobsters among others)? What are the advantages and disadvantages of that system?
• What steps (diet, exercise, posture) can people take to maintain the health of their skeletons?

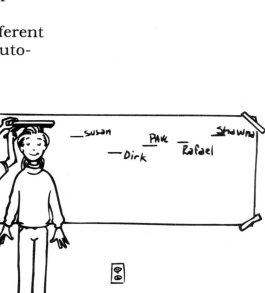

Skeleton Diagram

Scientists and doctors use special names for bones. Many bones also have everyday names. Study the skeleton and try to list the popular names for the following bones:

- carpals
- clavicle
- cranium
- femur
- mandible
- patella
- pelvis
- radius
- sternum
- tarsals
- tibia
- vertebrae
- zygoma

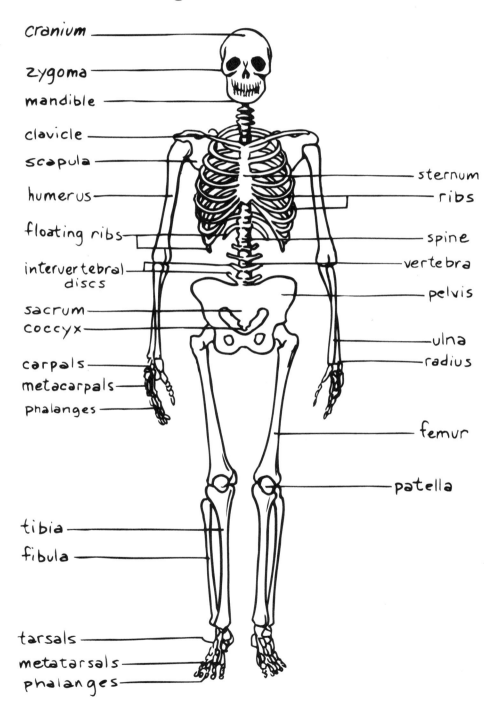

SPEECH

Speaking seems like a simple activity. But saying "Hello" or "The rain in Spain" is more complex than one might think at first. By studying phonetics, students may begin to understand that scientific observation can lead to a sense of wonder.

DIRECTIONS:
1. Divide the class into groups of two or three.
2. Give each group a copy of the Speech Sounds to Analyze handout, next page.
3. By observation, have students list which consonants are formed using the lips (b, f, m, p, v, w).
4. Optional: Have students give fuller descriptions of the forming of each consonant. This will include:
• <u>use of tongue</u>: for example, with d, the tongue presses against the top of the mouth.
• <u>use of teeth</u>: for example, with v, the upper teeth press down on the lower lip.
• <u>force of breath</u>: compare b (soft) with p (explosive)
5. Share the results.

EXTENSION:
Have students carry on similar research analyzing the production of vowel sounds and diphthongs. In addition to making traditional reports, students might create ABC picture books teaching younger children about speech sounds. Related topics for library research include:
• How do ventriloquists use phonetics to fool people?
• How do English speech sounds compare with those in French (or another language)?
• Can apes make sounds that resemble human speech sounds?

Speech Sounds to Analyze

consonants		vowels	
b	bike	a	ate
c	cat	a	ant
d	dog	a	are
f	far	e	eat
g	good	e	egg
h	hat	i	ice
j	jump	i	igloo
k	kick	o	oak
l	luck	o	of
m	map	o	off
n	nap	o	or
p	pay	u	us
q	quick	u	urge
r	run		
s	sun	**diphthongs**	
s	pleasure	br	brick
t	take	ch	chin
v	vine	ng	sing
w	wish	oi	oil
x	Xerox	oo	mood
y	yell	ou	out
z	zoo	sh	shop
		th	that
		th	the
		th	thin
		wh	when

SPEED

Science often involves measuring the speed of things. This knowledge can have practical importance, for example, determining the safe deployment rate of air bags.

MATERIALS:
- measuring stick or tape
- watch with second hand (or stop watch)

DIRECTIONS:
1. Ahead of time, measure a flat track of about 30 feet (10 meters). The number is arbitrary.
2. Explain that speed is a measure of distance traveled in a given time. For example, the speed of light is about 186,000 miles per second (301,000 kilometers per second). "Per" means "during a."
3. Ask students to estimate their normal walking speed. Then have them determine the speed experimentally. This will involve three steps:

   A. Time how long it takes (in seconds) to walk the measured distance. For example, a student might cover 30 feet in 7 seconds. For accuracy, do two trials and average the results.

   B. Divide the time (7 seconds) into 3600 seconds (the number of seconds in an hour). In our example:
   3600/7 = 514

   C. Multiply the experimental distance by the answer in Step B (514) to calculate the distance that would be covered in an hour at the rate of 30 feet/7 seconds. In our example
   30 feet x 514 = 15,420 feet/hour

   D. Divide the answer in Step C by 5,280 (the number of feet in a mile) to get the distance in terms of miles. In our example:
   15,420/5,280 = 2.92 miles

Thus, the walking speed is 2.92 miles/hour.

EXTENSION:
Have students measure the speeds of other moving bodies, such as a crawling baby, a tossed ball, or a slow-moving animal (snail). For a related library research project, use the handout on the next page.

What's your hurry?

Speeds to Investigate

Do library research to find the speed of the following:

baseball pitch
bicycle
bird
blood
bullet
crocodile
Earth traveling around the sun
film moving through a projector
horse
insect flying
message from toe to brain
ocean waves
river flowing
shark
skier
snail
snake striking
sneeze
submarine
swimmer
telephone call
train
wind

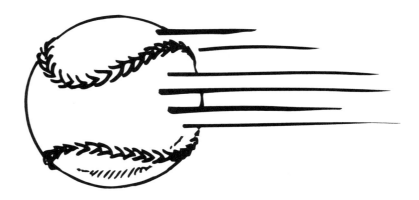

SUPERSTITIONS

Although superstitions are, by definition, unscientific, such beliefs can be studied scientifically.

DIRECTIONS:
1. Review what a superstition is: a belief in the magical power of a thing, idea, or action. An example is the belief that Friday the 13th is an unlucky day. You might ask students to hypothesize why a specific superstition came into existence. For example, walking under a ladder could lead to an injury.
2. Choose a superstition. You'll find a list on the next page.
3. Ask students to devise ways to determine the validity of the superstition. For example, one could check the unluckiness of Friday the 13th by reading through past newspapers and seeing if more accidents occur on that day than on other Fridays.
4. Share the investigative methods. Discuss the difficulties involved in evaluating concepts such as "good luck" or "bad luck."
5. Optional: Have students carry out the research projects that they devised above.

EXTENSION:
Have students use polling to assess the level of belief in superstitions. For example, your class might poll students in different grades to learn whether younger or older students are more superstitious. For help in preparing a data-collecting poll sheet, use the material on the next page.

Taking a Poll

A poll is a way to gather information about what people think. Usually, you will ask many people the same questions, and then analyze their responses.

Ahead of time, prepare a sheet for collecting the answers. Here is a sample. Your version might be very different.

After you prepare your poll sheet, you can make a copy for each person you talk to, or you can keep one copy in your notebook and write the answers on separate pieces of paper.

Biographical data
For each person polled, give this information:
- gender (male or female)
- age
- education level

Instructions for each person in the poll:
1. How superstitious are you? Rate yourself from 0 ("I'm not at all superstitious") to 10 (I'm very superstitious").

2. Do you carry a lucky charm—yes or no?

3. Rate each of the following superstitions from 0 ("It's total nonsense") to 10 ("It's absolutely real").
   A. The number 13 is unlucky.
   B. Step on a crack, break your mother's back.
   C. Spilled salt brings bad luck.
   D. The number 7 is a lucky number.
   E. It's bad luck to cross the shadow of a black cat.
   F. If you break a mirror, it's bad luck.
   G. Carrying a four-leaf clover brings good luck.
   H. Knocking on wood brings good luck.
   I. It's bad luck to open an umbrella indoors.

4. Are people today more superstitious, less superstitious, or about the same as people who lived long ago?

SWEAT

Antiperspirant ads try to convince us that sweat is bad. This lesson explores the other side of the story.

MATERIALS:
• small, slightly damp cloths (socks will do)

DIRECTIONS:
1. Ask students to write down everything they know about sweat. Pose questions such as:
• When does it occur?
• What "good" does sweating do in terms of comfort or survival?
2. Tell the students they're going to do a sweating simulation. Divide the class into small groups. Give each group a damp cloth. The cloth will represent the skin when sweaty.
3. Each group member should touch the cloth to assess its temperature.
4. Have one member in the group twirl the sock at a distance from the other students. Then have each group member feel the sock again.
5. Discuss why the socks are cooler after twirling. See the notes in the margin. Relate this cooling to what happens to a person on a hot day.

Cooling by Evaporation
• The human body can be thought of as a factory that runs on food energy. This factory turns out work, such as thinking, running, singing, and drawing.
• Like all factories, the body must be kept at the proper temperature. For most people, this is about 98.6 F (37 C).
• When the air temperature rises to an uncomfortable level, thousands of sweat glands produce sweat (a salty solution). On the average, there are 650 sweat glands per square inch (6.5/ sq. cm.)
• As sweat turns to vapor, it takes heat from the body, and combats overheating.

EXTENSION:
Use the handout on the next page to help students understand why high humidity days are more unpleasant than low humidity days. Related topics for library research include:
• Why is sweat salty?
• Do other animals sweat?
• What causes body odor?
• Why do people sweat when nervous?
• What steps can prevent heat stroke?
• What first aid steps are called for if someone suffers heat stroke?

Answers for next page:
1. No 2. Yes 3. Yes

Avoiding Heatstroke

The amount of water vapor in the air is measured by "relative humidity," or "humidity" for short. When the humidity is 0%, the air feels dry. At 100%, there's often rain.

Evaporation occurs slower when air contains a lot of moisture. If the actual temperature stays the same while the humidity rises, a person feels warmer. More important, the body's natural cooling system works less well. The body can overheat. This can be more serious than just feeling hot. Overheating causes heat stroke, which can be deadly.

The Heat Index allows you to determine dangerous combinations of heat and humidity. When the number is 105F or more, a person is in danger of heat stroke. Study the Heat Index, then answer the questions below.

Heat Index Chart

The % at the top of columns 2 through 7 represent humidity readings. The numbers below the humidity represent what the temperature feels like at that humidity. For example, if the actual temperature is 80 (column 1), at 100% humidity (column 7) it feels like 91.

Col 1	Col 2	Col 3	Col 4	Col 5	Col 6	Col 7
	0%	20%	40%	60%	80%	100%
Actual Temp.	Temperature a Person Feels (degrees Fahrenheit)					
70	64	66	68	70	71	72
80	73	77	79	82	86	91
90	83	87	93	100	113	122
110	91	99	110	120	144	150

Questions

1. Look in Column 2. If the humidity is 0%, are you likely to have heat stroke when the actual temperature (given in Column 1) is 110F?
2. If the actual temperature is 110F, is there a heat stroke danger if the humidity is 40%?
3. When the actual temperature is 90F, are you in danger of heat stroke if the humidity is 80%?

TASTE

In the past, taste was a tool for survival, enabling people to avoid bad food. Today, it's viewed mainly as a source of pleasure ("taste treats"). But in the classroom, taste can be the focus of serious research.

MATERIALS:
- jars of different baby food fruits with the same texture and color
- a small paper plate for each student
- a plastic spoon or stirring stick for each student

DIRECTIONS:
1. In their journals, have students tell what they know about taste.
2. Discuss their ideas, adding information as needed from the Taste Basics box.
3. Guide students through the following test designed to isolate taste from other sensory inputs, such as color and texture:
- Have students number two or more areas on their plates.
- Dab a small amount of the first baby food onto section 1, a small amount of the next type onto section 2, and so on. Make sure students do not see the labels on the jars.
- While holding their noses to block smell clues, students should taste and try to identify each fruit.
4. Go over the results.
5. Optional: Have students try similar taste tests, for example, comparing different sodas or juices, or whole fruits (apple, pear) without using visual or olfactory (smell) clues.

EXTENSION:
Different areas of the tongue are sensitive to different tastes. (See diagram, next page.)
Have students at home map their taste buds using the procedures outlined on the next page. Related library research topics are:
- How does gustation change as we age?
- Do animals have the same taste sensations as people do?

Taste Basics
- The scientific word for taste is *gustation*.
- There are about 3,000 taste buds on the tongue.
- When chemicals in foods dissolve in saliva and reach the taste buds we taste something.
- There are only four tastes: sweet, salty, sour, and bitter.
- We can tell one food from another largely with the help of the other senses.
- Smell plays a big role in food recognition, as is clear when we have a stuffy nose. Because the nose and mouth connect at the top of the throat, we can smell food even as we chew it.
- Texture, temperature, color, and sound also let us identify foods.

Ten-Minute Real World Science ©1997 Monday Morning Books, Inc.

Tongue Mapping

The tongue is the human organ for taste. It can recognize four tastes: sweet, salty, sour, and bitter. Certain areas of the tongue are more sensitive to different tastes. You can map these areas in the following way. You will need a partner.

Materials:
- small cups containing the following liquids and labeled so as to distinguish them:
    sugar water (sweet)
    salt water (salty)
    unsweetened cocoa in water (bitter)
    vinegar or unsweetened grapefruit juice (sour)
    plain water
- cotton swabs
- drawing materials

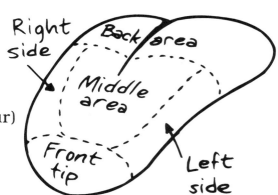

1. Draw an outline of a tongue. Divide it into sections as shown in the illustration.

2. Dip a swab into one of the liquids without telling your partner which it is.

3. While your partner holds his or her nose, touch each section of your partner's tongue. Be careful when touching the back area. It may cause a gag reflex. When your partner reports the taste, label the position on the tongue map. For example, if you touch the sugar water swab in the front and your partner tastes something *sweet*, label that section sweet. Continue through all the parts of the tongue.

4. When done with the first liquid, have your partner rinse with plain water, then test the next item. Continue this way through all the liquids.

5. If you like, have your partner test you, but use separate swabs.

TEETH

Aristotle taught that males have more teeth than females. Centuries later, Bertrand Russell said that Aristotle could have avoided his error by "asking Mrs. Aristotle to open her mouth." This comment underscores the value of observation in science.

DIRECTIONS:
1. In their journals, have students briefly describe what they already know about their teeth:
• How many are there?
• How many kinds are there?
2. Explain that classifying means dividing a group of items into subgroups. This involves finding similarities in shape or function or both.
3. Divide the class into pairs. Ask students to study the partner's teeth—but NOT touch them—and identify several types. Students should give each type a descriptive name, for example, "slicers" or "grinders." If time permits, the observation might include sketching each type of tooth. Note: If your students would find looking into a classmate's mouth too distracting, they can do this activity with a mirror or at home with a parent's help.
4. Have students discuss their findings.

Hey, you have canine teeth just like me!

EXTENSION:
Have students use the Guide to Human Teeth, next page, to learn the scientific name of each tooth and the features of teeth in general. Related library research topics are:
• the function of each kind of teeth
• oral hygiene
• comparing human and nonhuman teeth

Teeth Basics
• During a normal life, people get two sets of teeth.
• The "baby" or "milk" teeth are present at birth but don't begin to appear until about seven months. Two more years pass before all 20 baby teeth grow in.
• At 6 or 7, the baby teeth lose their roots and are pushed out by the second set, called adult teeth.
• A full set of adult teeth consists of 16 upper and 16 lower teeth, 32 in all.
• A tooth has a visible upper part, the crown, covered by white enamel, and a hidden part, the root, set in the jaw. Under the enamel is a hard material (dentine). In the center are blood vessels and nerves.
• The teeth of plant-eating mammals, like cows, grow as the tops wear out.

Guide to Human Teeth

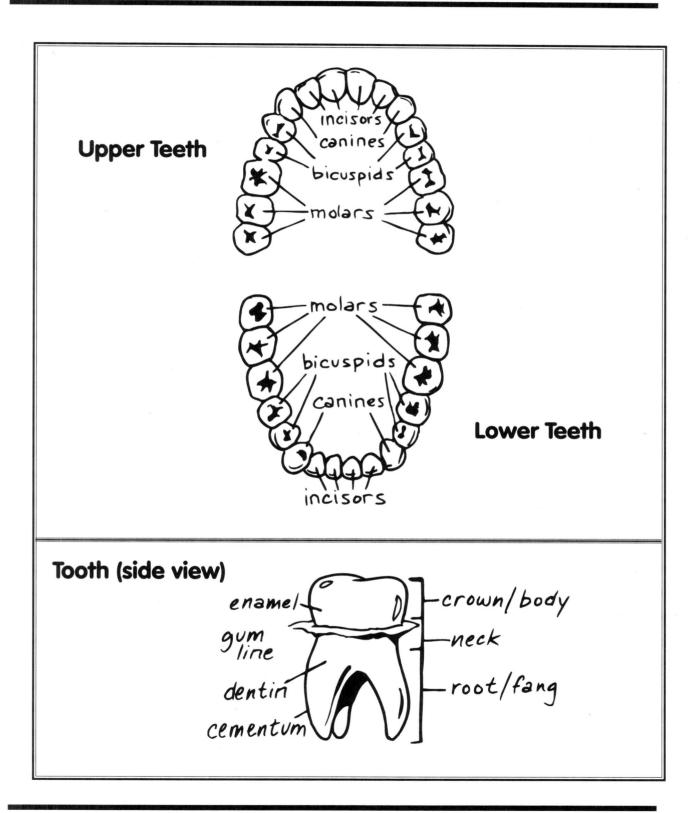

THINKING MACHINE

To simulate human thought, computer scientists try to analyze the steps experts go through when carrying out activities, such as playing chess or diagnosing illnesses. Students can experience this effort by describing in plain English the steps a computer could follow in handling a task. This is what advanced scientists do before translating the steps into computer language.

DIRECTIONS:
1. Make sure students know how to play "Ghost," a game for two or more players:
• To start, someone says a letter.
• Players take turns adding letters, trying NOT to end up spelling a word.
• A player who unintentionally spells a word gets a *g* the first time, then an *h*, and so on. A player loses by getting stuck with all the letters in "ghost."
2. Explain that some computer scientists try to create "artificial intelligence" by listing the steps of a task. In effect, this is giving directions to a computer.
3. To prepare students for writing their own artificial intelligence "program," give them copies of the Script for Ghost, next page.
4. Assign one student to play the role of the computer while another student plays the human opponent.
5. Read the directions aloud while the "computer" and human play.
6. Discuss the results. Can students think of ways to make the steps clearer? Are there any steps that should be added?

EXTENSION:
Have students choose a simple game and write directions to teach a computer how to play it. Possible games include:
• Tic-tac-toe
• Hang Man
• Checkers
• Dominoes

Sample Game of Ghost
Player 1 says "j"
Player 2 says "a" (j-a)
Player 1: says "r" (j-a-r)
Player 2: "That spells a word!"

Note 1: If player 1 had added a "c" instead of an "r" the game would have continued.

Note 2: If a player adds a letter that makes an impossible combination, the other player can challenge. For example, if player 1 had added a second "a" (jaa), player 2 might have challenged because no English word starts with the letters "jaa."

I want to play, too!

Script for Ghost: Computer vs. Human

The actor who plays the computer should follow the directions after COMPUTER, and then read aloud the words in **boldface** type. The actor who plays the human should follow the directions after the word HUMAN.

1. COMPUTER: (Say any letter of the alphabet—a to z—and then say): **OK, human, please add a letter.**
    HUMAN: (Add a letter so that you spell the beginning of any word that will have three or more letters.)

2. COMPUTER: (If no possible word can be spelled by adding another letter, go to Step 9. If you must complete a word, go to step 8. Otherwise, add a letter and say): **Your turn. Add a letter.**
    HUMAN: (Add a letter that could be part of a word, but try not to spell a complete word.)

3. COMPUTER: (If the human has <u>not</u> spelled a word, go to Step 6. If the human <u>did</u> spell a word say): **You spelled _____. I win. To play again, say "Play." To quit, say "Quit."**
    HUMAN: (Say either "Play" or "Quit.")

4. COMPUTER: (If the human says "Play" go to Step 1. If the human says ""Quit," go to Step 5.)

5. COMPUTER: **Goodbye.**

6. COMPUTER: (If you can add a letter without spelling a word, say it. Then read the boldface words in Step 2. Otherwise, go to Step 7.)

7. COMPUTER: (If you are forced to add a letter that completes a word, go to step 8. If there is no possible letter to add, go to Step 9.)

8. COMPUTER: **I am forced to complete the word _____. So you win. To play again, say "Play." To quit say "Quit."**
    HUMAN: (Say "Play" or "Quit.")
    COMPUTER: (If the human says "Play," go to Step 1. Otherwise, go to Step 5.)

9. COMPUTER: **There is no possible letter that follows the letters, so I win. To play again, say "Play." To quit, say "Quit."** (If the player says "Play," go to Step 1. Otherwise, go to Step 5.)

TOUCH

The following activity shows the value of developing all the senses, and also can serve as an introduction to non-visual science observations, for example, the use of sonar in mapping the ocean floor.

MATERIALS:
- several grocery bags or pillow cases
- small objects <u>without sharp edges</u> (see next page)

DIRECTIONS:
1. Ahead of time, fill the bags with four or five objects. Include those with familiar shapes, for example, a comb and a key; and those that will be more difficult to identify by touch, for example, a popcorn kernel or a postage stamp.
2. Explain to the students that scientists must often work with limited information. A well-known example is studying distant stars. The following activity simulates this situation.
3. Divide the class into the same number of groups as there are bags. Give each group a bag, but instruct the students NOT to look inside.
4. Tell students to take turns putting their hands inside the bags. Using only the sense of touch, they should not only try to identify the objects, but also describe them in detail, listing such features as shape, size, and texture. For example, if they identify a button, they might report on its size, its number of holes, and the material it's made from.
5. Optional: If time permits, rotate the bags among the groups.
6. Reveal the contents and discuss reasons that led to correct and incorrect guesses.

EXTENSION:
Do the same activity using the sense of hearing. Place several objects in a sealed shoe box, for example, a marble, a handful of paper clips, and a couple of pencils. Allow the students to shake or otherwise manipulate the box without looking inside. The challenge is to figure out the contents by the sounds that the objects make.

Touch and Tell Items

- adhesive tape on a roll
- aspirin bottle with safety cap
- bar of soap
- bead with hole for stringing
- business card
- carrot
- chess piece
- coin
- combination lock
- computer diskette
- cover from a slick magazine
- envelope
- eyedropper
- eyeglasses (without lenses)
- feather
- funnel
- golf ball
- greeting card
- key
- magnifying glass
- measuring cup
- paper clip
- pencil sharpener
- Pez dispenser
- phonograph record (45 rpm)
- photographic slide
- piece of yarn
- plastic bag
- plastic picnic spoon
- postage stamp
- potholder
- refrigerator letter with magnet
- shoehorn
- shuttlecock (from badminton game)
- rubber band
- string bean
- washer
- watch band
- whistle

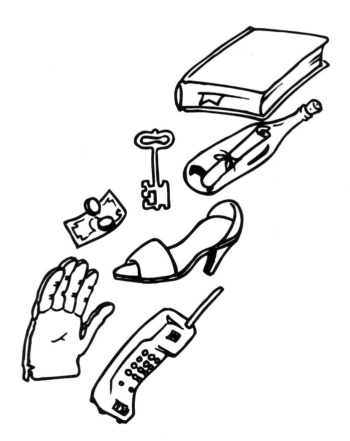

WATER

Water is both commonplace and important. That makes it a wonderful substance for research.

MATERIALS:
- ice cube
- 3 clear glasses filled with water (one very narrow)
- pencil or stick

DIRECTIONS:
1. Have students write a description of water for someone who has never seen it. For starters, brainstorm expressions such as: *clear, wet, drinkable, takes the shape of its container.*
2. Set up three water stations. Have students observe and try to explain what they see:
- ICE FLOATS: Float an ice cube. Have students estimate the fraction of ice above the water line. [Facts: About 1/10 of the ice is above the water line. In salt water it would be about 1/7. Hence the expression, "the tip of the iceberg." When water freezes, it expands. This makes ice less dense than water. Anything that is less dense than water will float.]
- LIGHT BENDS (REFRACTS): Place a pencil in a glass of water. Have students describe, draw, and explain what they see. [Fact: When light moves from one medium to another, its speed changes and so does its path.]
- MOLECULES ATTRACT: Carefully fill the narrow glass so that the water rises above the edge! [Fact: Water molecules bond so strongly, they form a kind of skin. A drop of detergent will chemically break the "surface tension" and allow the water to flow over the edge.]

EXTENSION:
For other water experiments, see the next page. Related library research questions are:
- Why is the ocean salty?
- How do plants use water?
- What causes the sense of thirst?
- How do the kidneys recycle water?
- What causes rain?
- How do cities purify drinking water?

Water Basics
- Water and ice cover 70% of the earth.
- A water molecule (the tiniest bit of water) is made of 1 oxygen and 2 hydrogen atoms (H_2O).
- Many things dissolve in water. That's why water is important in making chemicals and for cleaning things.
- Water stays warm. This makes it useful for heating devices, such as hot water bottles.
- Using solar power, plants turn water and air into food products.
- Water molecules are always moving. That's why a drop of food color will eventually stain all the water in a glass.

Water Experiments

Filtering Water
Prepare a large container of muddy water. Then try to remove the mud in these ways. DO NOT DRINK THE WATER.
• Pour the water through a coffee filter set in a cone or funnel. Try the same activity using two filters. What's the difference in water quality and in time of filtration?
• Do the same experiment using other materials for filters, such as plain typing paper, newsprint, and cotton cloth.
• Punch or cut small openings in the bottom of an empty juice can. Fill the can with gravel or with sand and then gravel. Place it in a bucket. Pour the muddy water into the can. Compare the water with that from the other experiment.

Ice
Pour equal amounts of various liquids into separate metal or plastic freezer trays. In addition to plain water try things like:
• soda water   • juice   • salt water   • soapy water

Place the liquids in a freezer. Check every half hour. What happens? Note: NEVER FREEZE WATER IN A GLASS CONTAINER OR IN A CLOSED CONTAINER. As it freezes, water expands and can crack containers. This is why water pipes sometimes burst in cold weather.

Clean Up
Prepare three samples of stained fabric. Try:
• organic stains (catsup, grass, juice, vegetable oil)
• dirt stains

Wash one sample in plain water, another in soapy water, and the third in detergent water. (Detergents are chemically different from soaps).
• Compare the cleaning solutions for each stain.
• As a variation, see if water temperature makes a difference.

Currents
Fill two glasses, one with hot water, the other with cold water. Let the glasses stand. Then carefully drop a single drop of food coloring into each glass. Describe the results.
• Does the coloring spread equally fast in both glasses?
• How long does it take for the color to become evenly spread?

WAVES

High-tech equipment is often used to study waves. But if you have a jump rope handy, your students can study this topic, which relates to many phenomena, including sound, light, earthquakes, and thinking.

MATERIALS
- rope (a jump rope will do)
- optional: a pan of water

DIRECTIONS:
1. Have students think about waves they've seen at the beach and in the bathtub, and then write a description and definition of the phenomenon.
2. Using the rope, have two students create waves at the front of the room, varying the frequency. The onlookers should draw what they see.
3. Use the Wave Diagram, next page, and information in the box to review the basic concepts.
4. Optional: Give students a chance to explore wave action by rocking a pan of water from side to side, or by dropping pebbles into it. They should record their observations.

EXTENSION:
On the playground, have students crouch in a circle and perform "the wave," by standing up and then crouching in sequence. Later, discuss the fact that the wave moves along, but that students stay in place.
Related library research topics are:
- What are sound waves?
- What are radio waves?
- What are brain waves?
- How do seismic waves cause damage?
- How does wave action move food through the digestive system?

Wave Basics
- A wave is an up-and-down, side-to-side, or back-and-forth vibration of moving energy.
- Types of waves include: sound waves, light waves, brain waves, and seismic (earthquake) waves.
- Waves often involve the motion of a material, such as water or air.
- Although a wave's energy may travel great distances, such as across an ocean, the vibrating material itself stays in the local area. Thus, a dyed patch of water will stay offshore as waves pass through it. Likewise, in a "people wave," the wave moves around the stadium, but the people stay near their seats.
- Waves are cyclic. The number of cycles per unit time is the frequency. High frequency waves transport more energy than low frequency waves.

Wave Diagram

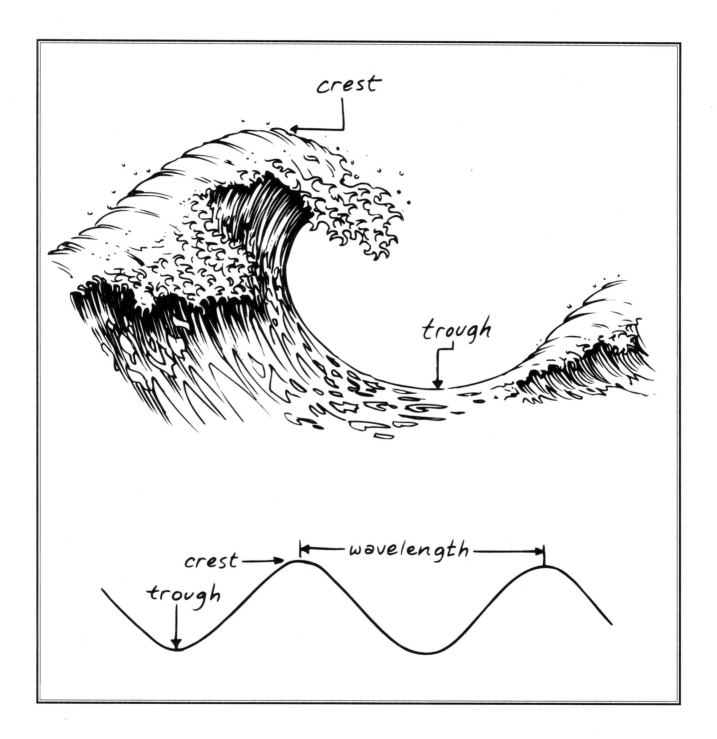

SKILLS: CLASSIFYING

To cope with the seemingly infinite number of individual things and events, scientists work out classification systems. This work requires both knowledge and critical thinking.

Things to Classify
blood vessels
bones
earthquakes
elements
foods
hurricanes
illnesses
rocks
soils
stars
words

DIRECTIONS:
1. Explain that classifying means putting things into groups according to important similarities, such as structure and behavior. For example, although birds and bees both fly, their structures and life cycles are so different, they are placed in different groups or classifications.
2. Divide students into teams.
3. On the board, list three animals, two of which are related. For example: whale, shark, goldfish.
4. Have students use their knowledge and critical thinking skills to identify which of the two animals go together and to explain why. For example, although whales and sharks are both much bigger than goldfish, sharks and goldfish are related because they both hatch from eggs and breathe using gills. (See the next page for more examples.)
   Note that scientists have often disagreed about how to classify animals or other things. What's important in this exercise is that students have reasons for their groupings.
5. Have students share their views. Then go over the classifications used by professional scientists.

EXTENSION:
Have students choose animals, and then write reports on how the animals are classified. The reports should list related animals. Later, students might report on the systems used to classify other things, such as those in the box.

72

Ten-Minute Real World Science ©1997 Monday Morning Books, Inc.

Animals to Classify

bats, eagles, robins: Eagles and robins are both birds, born from eggs. Bats are mammals, which are born live.

bees, butterflies, spiders: Bees and butterflies are insects, with six legs and three main body parts. Spiders are arachnids, eight-legged creatures whose bodies are divided into two parts.

bears, cows, sheep: Cows and sheep are vegetarians. They eat grass and plants. Bears are omnivorous. They eat meat and plant food.

dogs, house cats, lions: House cats and lions are in the cat family, whose members have sensitive whiskers and sharp claws. Dogs are related to foxes and wolves.

crabs, frogs, salamanders: Frogs and salamanders are amphibians, meaning that they spend the early part of their lives in water, and later live on land. They have four legs. Crabs are shellfish, related to lobsters, and have eight legs.

fleas, grasshoppers, kangaroos: Fleas and grasshoppers are both insects, with six legs. They grow from eggs. Kangaroos are mammals, whose young are born alive.

fireflies, houseflies, ladybugs: Fireflies (also known as lightning bugs) and ladybugs are beetles, a kind of insect that has hard wings and biting mouths. Houseflies are a kind of insect that has soft wings.

alligators, snake, worms: Alligators and snakes are reptiles, animals that have skeletons and are covered with scales. Worms have soft bodies, and often live underground or as parasites.

SKILLS: Drawing

Scientists often express themselves artistically. Pasteur painted. Einstein played the violin. Carl Sagan wrote books and scripts. Drawing is the art most immediately relevant to science. Researchers draw what they observe and diagram their experimental set-ups. This graphic work doesn't require genius, but it does demand skills and practice.

DIRECTIONS:
1. Give each student a copy of the Drawing for Scientists handout, next page. Go over the tips.
2. Regularly challenge students to draw from observation. Include fixed subjects, such as the structure of a chicken bone; and also changing subjects, such as the day-by-day growth of mold on a slice of bread.
3. Optional: For subjects that can't be observed firsthand, such as the splash of a water drop caught by a stroboscopic flash, students can draw from photographs,
4. Have students share their work in small groups or by posting it. Make sure they understand two things about evaluation:
• Clarity and accuracy are the goals.
• Progress should be measured by examining one's own work, not the work of others.

EXTENSION:
Have students include original drawings with all their science projects.

Fixed Subjects
animal (fish, turtle)
ear
eye
flower
fruit exterior
fruit interior
hand
insects
leaves
magnetic field
mirror image
moths
rocks
snowflake
teeth
tongue
walnut exterior

Changing Subjects
a banana ripening
blood circulating
lips forming sounds
pendulum swinging
thrown ball (its arc)
a top spinning
waves in water

Drawing for Scientists

You don't need to be a great artist to make useful scientific drawings. You do need to notice details, and you do need to practice. Here are some tips.

Tip 1. Look before you draw. The key to accurate drawing is the eye, not the hand. Take time to study the subject carefully. When drawing a small subject, get close up, or use a magnifying class or a microscope.

Tip 2. Use a pencil. This way you can easily make corrections. When you're satisfied with your drawing, go over it in ink, or add color.

Tip 3. Draw big. A larger drawing makes it easier for you to capture the details.

Tip 4. From time to time, stop drawing and compare the subject with your drawing. When you see differences, make corrections.

Tip 5. Create more than one drawing, if needed. For example, when drawing a leaf, you might want to draw both sides. When drawing an insect, you might want a top view, a bottom view, a side view, and a front view. When drawing a subject that changes, for example, a plant growing or the waxing and waning of the moon, you'll need a series of drawings made at different times.

Tip 6. Label each finished drawing. Include your name, the name of the subject, the date the drawing was made and the location, the subject's actual size (if known), and the names of the parts (if known).

Tip 7. Be patient. Like learning to ride a bike or read a book, learning to make scientific drawings takes practice. The first drawings you make may not please you, but they will be steps toward success.

SKILLS: EXPERIMENTING

Modern science came about largely because of the development of the experimental method. The word *experiment* relates to *experience*. Both words trace back to an ancient expression meaning "to try." Rather than be told that something happens, scientists want to try it themselves. For students, this kind of real-world science is truly learning by doing.

DIRECTIONS:
1. Define "experiment." See the box in the margin.
2. Review the steps scientists follow in carrying out an experiment. These typically are:
- **Choose a problem** to work on, for example, learning to spell words.
- **State the problem as a question,** such as: "Does writing play a role in learning to spell?" One problem might lead to many questions. The challenge is to pick the most interesting or important questions.
- **Develop an idea for a way to answer the question.** This step often involves trial and error, and can take a long time.
- **Put the experiment together.** This includes finding or creating materials.
- **Try the experiment.** Often, the first experiment won't work and the scientist must make changes in the plan, or even start over.
- **Report the results.** See the model, next page. A detailed report allows other scientists to think about and try the experiment themselves. They may find weaknesses in the experiment that will require a new experiment.
- **Explore related problems.** Discoveries made by one experiment often lead to other studies.
3. Demonstrate a few experiments. You might do this by following directions in books of experiments (see Resources) or by having guest scientists share experiments with your class.

EXTENSION:
Encourage students to create their own experiments. Suggest that they work in groups. Group brainstorming often leads to the generation of many ideas. Besides, some of history's important experiments were done by teams.

Experiment Basics
- An experiment is an observation in which the situation is controlled.
- This control allows the experimenter to focus on a single event, the "independent variable."
- Control also makes it possible to repeat the observation.
- Repetition is a key strategy for overcoming error and gaining confidence in the finding.
- The ability to repeat an observation also means that other scientists will have the opportunity to confirm the finding.

Sample Experimental Report

Using Writing to Learn Spelling Words

Question: Will people learn to spell better if they read <u>and</u> write the words instead of just reading the words?

Materials: Ten nonsense words of 6 to 8 letters used were:
brubtley	feearp	Kwengop	wuyfle	selnool
klivsa	voodstim	nopstell	umgwas	destrig

Method:
1. The subjects were 20 sixth graders divided randomly into two groups. Group 1 was the "Read only" group. Group 2 was the "Read and write" group.

2. Each subject got a page with the 10 nonsense words to study for 10 minutes. Nonsense words were used to make sure that the subjects did not already knew how to spell any of the words.

3. Group 1 was told to read the words only. Group 2 was told to read the words and also to write each one at least three times.

4. After 10 minutes, the experimenter collected the word lists.

5. The experimenter then read aloud the words twice, allowing subjects 15 seconds to write each word.

6. One day later, subjects were retested.

Results:
On the first test, Group 1 subjects averaged 4 words correct; Group 2 subjects averaged 6 words correct. On the next test, Group 1 averaged 3 words correct; Group 2 averaged 5 words correct.

Conclusion:
Writing seems to help in learning spelling words.

Future experiments:
Some Group 1 subjects read the words aloud, others silently. The next experiment will compare the two kinds of reading. Later, the main experiment will be done using real words.

SKILLS: FACT CHECKING

Although scientists aim to find new knowledge, they also need to know how to find "old" facts. That's because innovation almost always rests on past discoveries.

DIRECTIONS:
1. Give each student a "Facts to Check" handout.
2. Have students use their background knowledge to answer each question "True," "False," or "?" ("I don't know").
3. Discuss sources students could use to find or check the facts. Possibilities include:
• experts
• books (including reference books)
• the Internet
• direct observation

Fact Checking Report
Fact to check: "Diamonds are the hardest mineral."
<u>This is true.</u>
Source 1: *The Concise Columbia Encyclopedia* (Avon, 1983). Page 231.
Source 2: *The Penguin Book of the Physical World* (Penguin, 1976). Article 18.

EXTENSION:
Have students, individually or as teams, do research scavenger hunts to check their answers. For each answer, they should give:
• the name of the resource
• relevant information, such as the publisher
For a greater challenge, require that students locate two sources for each fact. Share the results on a bulletin board or in a group discussion.

Facts to Check

Number a sheet of paper from 1 to 20. Next, read each statement below. Using your own knowledge, decide whether the statement is "true," or "false." If you don't have any idea, write "?" or "I don't know." Then do library research to check your answer.

1. Diamonds are the hardest mineral.
2. Insects have brains.
3. Planets can have more than one moon.
4. A flea can jump more than 5 times its length.
5. All spiders spin webs.
6. Dogs can hear sounds that people can't.
7. At sea level, sound travels at about 760 mph.
8. It takes light about 8 minutes to go from the sun to the earth.
9. The jaguar is the fastest land animal.
10. Only human beings have thumbs.
11. There are more calories in an apple than in an orange.
12. The sun gives out only visible light.
13. The earth is a perfect sphere (shape of a basketball).
14. In some places, the ocean is more than ten miles deep (16 kilometers).
15. Human beings live longer than any other animal.
16. The brain contains about one million neurons.
17. It takes sunlight three minutes to reach the earth.
18. Ice floats because it's less dense than water.
19. A nanosecond is one billionth of a second.
20. Glass is a fluid.

SKILLS: MEASURING

Scientists have a passion for exact measurement. That's because it often leads to great discoveries.

MATERIALS:
- measuring stick
- scale
- thermometer
- measuring cups

DIRECTIONS:
1. Explain that accurate measurement is a key step in scientific research.
2. Choose something to measure, for example, the area of the classroom. (See more suggestions in the margin.)
3. Have each student or a team of students make the measurement several times and record the results.
4. Have students or teams compare their measurements. Discuss difficulties encountered in the work. If there are discrepancies, ask students to remeasure.

EXTENSION:
Have students measure a changing subject, such as the growth of a fingernail. Related research can focus on measurement breakthroughs, such as how scientists determined the speed of light. Students can read about some of these topics, but might also correspond with experts, for example, contacting a biologist about measuring the speed of a sneeze. For a list of topics, see the next page.

Things to Measure
Length
- animal in classroom
- flag
- pencil

Area
- paper money
- teacher's desk
- textbook

Circumference
- coin
- doorknob
- finger
- head
- tin can

Weight
- coin
- eraser
- index card
- quart/liter of water
- textbook

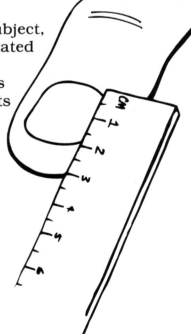

How Do Scientists Measure...?

age
- of a tree
- of a rock
- of a fossil

energy
- in various foods
- in a gallon of gas
- of an earthquake

circumference
- of Earth
- of the sun

distance
- to the moon
- to the sun
- to the nearest star

duration
- of a dream
- of a year

height
- of Mount Everest
- of the moon's mountains

rate
- a hummingbird beats its wings
- a piano string vibrates at middle C
- blue light vibrates

size
- of an atom
- of bacteria
- of the biggest dinosaur
- of a distant star

speed
- of a pitched baseball
- of a hawk
- of light
- of sound
- of a shark
- of a sneeze
- of an avalanche
- of a snail

strength
- of a spider's web
- of a bridge

temperature of
- the sun
- Earth's core
- an atomic explosion
- the interior of a volcano

volume
- of blood in a human being
- of food a human stomach holds
- of rain in a storm (rainfall)
- of water in the ocean

volume (noise level)
- of a grasshopper's song
- of a jet engine
- of thunder

weight
- an ant can carry
- of an atom
- of a whale
- of helium
- of pollution entering the air daily
- Earth

SKILLS: MODEL MAKING

Creating models encourages careful observation. It also provides a means of sharing information.

MATERIALS:
• simple objects, such as walnuts, popcorn, lima beans (enough for each student or small groups to have one)
• modeling clay

DIRECTIONS:
1. Explain what a model is.
2. Give each student an object that can easily be modeled in clay. Examples include: a walnut, a lima bean, or a chicken bone.
3. Have students make clay models of their objects.
4. Share the models in small groups.

EXTENSION:
Have students make models of complex objects, such as:
• the brain
• the ear (outer portion)
• a flower
• a hand
• an insect
• the intestines
• a molecule
• the paw of dog or cat
• a tooth

Models can be based on eyewitness observations or by studying pictures. Use the finished models in a museum-style display. For a project that uses people to model the solar system, see the following page.

> **Model Basics**
> • A model is a copy of an object, such as a cell, or of a process, such as an earthquake.
> • Scientists create models to make abstract ideas easier to work with. Models are also used for instructional purposes.
> • Models can be made of many things. For example, toothpicks and marshmallows are fine for modeling molecules. Computer programs allow scientists to make digital models.
> • Models of objects can be larger or smaller than the original. Either way, they should be in proportion. That is, if one part of an object is twice the size of another part, the model should represent the parts in the same way.
> • A model that faithfully represents the proportions of an object is called a scale model.

Scale Model of the Solar System

Pluto, the most distant planet, is 100 times further away from the sun than the closest planet, Mercury. That's why it's tricky to accurately represent the solar system in a book.

A better way to illustrate the spatial relationships of the solar system is to make a large model outside. You'll need an path at least 100 feet (30 meters) long, plus 11 actors.

Each actor should carry a sign naming one of the planets, or the sun, or the moon.

The scale of this model is 1:195 billion. This means that 1 foot (.3 meters) in the model equals 195 billion feet (58.5 billion meters) in real space.

- The sun actor stands at one end of the area.
- Mercury stands 1 foot (.3 meters) from the sun.
- Venus stands 2 feet (.6 meters) from the sun.
- Earth stands 2.5 feet (.75 meters) from the sun.
- The moon stands next to the earth, practically touching.
- Mars stands 4 feet (1.2 meters) from the sun.
- Jupiter stands 13 feet (4 meters) from the sun.
- Saturn stands 24 feet (7 meters) from the sun.
- Uranus stands 48 feet (14 meters) from the sun.
- Neptune stands 75 feet (22 meters) from the sun.
- Pluto stands 100 feet (30 meters) from the sun.

SKILLS: OBSERVING

We owe much scientific progress to the invention of microscopes, telescopes, microphones, x-rays, high-speed photography, spectrographs, and other observation tools. Nevertheless, the habit of careful observation can be developed without sophisticated equipment.

MATERIALS:
A set of four similar examples of something, such as:
- eggshells
- potatoes
- pieces of gravel
- plastic cups
- index cards
- socks
- pencils
- tiles

Label each item in the set, and find a subtle way to identify one of them, for example, by weight or by a unique mark.

DIRECTIONS:
1. In their journals, have students briefly describe the skills of observation. Then have them share their ideas, and share your own.
2. Tell students that they are going to carefully study one object, and then try to pick it out from a group of similar objects.
3. Have students observe and take notes on the identifiable item. They can use words and/or sketch the thing.
4. Later, place the object into the set of the three similar things.
5. Give students a chance to pick out the special member. They should be able to give a reason for their choice.
6. After everyone has made a pick, reveal the chosen item.
7. Optional: To sharpen observation skills, repeat this activity from time to time.

EXTENSION:
Make sure students have a wide variety of observation experiences. For observation tasks, use the list on the next page. In each case, students should record their observations in their science logs, and then share the results.

Observation Basics
- Observation means collecting information using the five senses: seeing, hearing, touching, tasting, or smelling.
- The successful scientific observer is usually looking for something specific, but is ready to notice the unexpected.
- When scientists observe something new, they almost always go back and observe it again to make sure they have seen it accurately.
- The task of describing something in words and pictures often leads to more accurate observation.
- Successful observation often requires a great deal of time.

Observation Experiences

There are many kinds of scientific observation. If possible, arrange to have at least one experience from each of the following groups.

Observe a process that happens quickly:
- eye blink
- lightning flash
- ball in flight
- insect in flight
- sneezing
- swallowing
- a top spinning
- a pendulum swinging

Observe a process that happens slowly:
- a seed sprouting
- water freezing
- iron rusting
- ice freezing
- a snail moving
- a fingernail growing
- fruit forming on a tree or vine
- a bird building a nest

Observe something with the aid of a tool:
- measuring stick
- camera
- scale
- magnifying glass
- microscope
- micrometer
- binoculars or telescope
- microscope
- chronometer (watch)
- prism
- thermometer
- compass

Observe a thing using a sense you never used before to study the thing, for example:
- use touch to observe a musical instrument while it's being played
- use hearing to observe your heartbeat
- use smell or touch to observe the soil
- use touch to observe your muscles
- use touch to observe the speech process

Observe something for one of these reasons:
- to confirm a fact, for example, that water expands when it freezes
- to test a hypothesis, for example, that kids are more likely to pick up trash than adults

SCIENTIST'S GUIDE

Science and scissors are related. Both words come from an ancient word meaning "to take apart." Scientists take apart things and actions to better understand them. This work focuses on rocks, flowers, stars, memory, lakes, sounds, and everything else in the universe. No wonder there are so many branches of science, from astronomy to zoology.

But here's something surprising: The many kinds of scientists have a lot in common. This Guide covers some of the ideas and methods that scientists share, whether they are learning about atoms or zebras.

Amazement: Are you filled with wonder when you see a rainbow, or watch a hummingbird fly straight up? Amazement pushes scientists to understand what they are seeing. When you have that feeling, pay attention. It might guide you to new knowledge.

Analyzing: When you take something apart to learn how it works, that is called analysis. It involves noticing details that are often overlooked. For example, if you analyze what happens when you talk, you'll notice that you're using your teeth, your tongue, your lips, the roof of your mouth, and your lungs.

Background knowledge: Scientists don't get ideas out of thin air. Often, their work begins by reading what other scientists have learned. For this reason, read as much as you can about a topic that interests you.

Drawing: One of the best ways to improve observation skills is to draw whatever is observed. You don't have to be a great artist. You just have to look carefully and then represent what you see. Most scientists sketch with a pencil, because it's easier to make changes.

Experiment: An experiment is a controlled observation. For example, to find out if smell affects taste, you could have people taste things while wearing nose clips. All the tasting would be done the same, controlled way.

Fact: A fact is something that can be proved to be true. For example, a scientist might say, "When you pour oil and water together, the oil will float on top of the water." Anyone who has oil and water can do an experiment and find out if oil really floats on water.

Hypothesis: A hypothesis is a trial explanation. For example, suppose you learn that people remember facts better if they go to sleep after studying them. You might guess that "the brain needs a quiet time in order to store information." That kind of guess, based on some facts, is a hypothesis. It usually leads to more research. For example, to test your hypothesis, you could ask people to study facts, and then sit in a quiet room without sleeping.

Laboratory (lab): In the movies, a scientist's laboratory is often a place with bubbling beakers and electrical sparks. But not all labs are like that. Some focus on people, not chemicals. Others are filled with computers. In fact, a lab is any place where a scientist does work. That shouldn't be surprising because the word *laboratory* comes from the word *labor*, "to work."

Law: A law in science is a statement that says one thing always causes another. For example, "Boyle's law" says that if you squeeze a gas, the temperature of the gas will increase.

Mistakes: Although no one likes to be wrong, mistakes are part of doing science. Even the most famous scientists make mistakes. For example, Louis Pasteur, the man who invented pasteurization, once thought he had discovered how to cure a silkworm disease. Many farmers followed his advice. But the silkworms continued to die. Pasteur had to admit he was wrong. Then he worked even harder to find the correct answer. You should not fear making mistakes, but you must always try to learn from them.

Observation: Scientists make discoveries by collecting information using the five senses—seeing, hearing, touching, smelling, tasting. Microscopes, telescopes, scales, prisms, thermometers, and other tools extend the senses. For example, the human eye can see only a few thousand stars, but with powerful telescopes, we can see millions more. Although tools are important in making observations, successful observation requires the following qualities:
• curiosity
• patience
• an open mind
• an interest in details

Precision: Most people know that it takes about a year for the Earth to go around the sun. But scientists have learned how to measure the time of this journey to a fraction of a second. Figuring out how to make accurate observations has led to many important discoveries.

Prediction: A prediction is a guess about what will happen under certain circumstances. For example, in an experiment about littering, you might predict that there will be less litter if there are more trash containers. You could then test that prediction.

Procedure: The steps that you follow when studying something is called your procedure. To help others understand your scientific work, you will need to describe the procedure you used. For example, if you do an experiment about growing vegetables from seeds, a report of your procedure might tell how deep you planted the seeds, and how often you watered them.

Publishing: Scientists don't learn things just for themselves. They share their discoveries with others. They do this partly because it's exciting to tell people about the universe. But in addition, they do it so that other scientists can try to observe the same things and learn if the observations are correct.

Questions: Scientists don't have all the answers. But they do have lots of questions. A question is often the starting point for scientific work. There are all sorts of questions. A few examples are:
- How is one thing like and different from another?
- How does something work?
- What do all these things have in common?
- How many of these things exist in one place?
- How big or how little is this thing?
- What causes this thing to happen?
- What are the parts of this thing?
- What are the steps of this process?

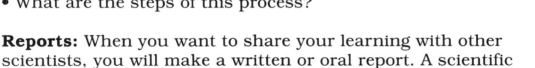

Reports: When you want to share your learning with other scientists, you will make a written or oral report. A scientific report will usually cover these points:
- Question: What puzzled you?
- Background: What have other scientists learned about the subject?
- Method and materials: How did you go about answering the question?
- Results: What did you learn?
- Interpretation: What do the results mean?
- Next steps: What is the next question you plan to answer?

Safety: Because scientists often study the unknown, their work can be dangerous. For example, some scientists who first studied radioactivity later became ill. They didn't know that radioactivity could be harmful. To protect yourself and those who work with you, it is important to be careful, especially with:
- chemicals
- breakable materials, such as glass
- heated materials
- unknown animals

When in doubt, smart scientists seek advice before trying an experiment.

Science Journal: Most scientists keep a journal in which they record their ideas, questions, observations, and experiments. When you make notes in your journal, start by writing the date. Also include diagrams of what you observed and sketches of any experiment that you set up.

Skepticism: If the most famous scientist in the world said, "Spiders have a language that is as complex as human language," most scientists would respond: "Prove it!" This show-me attitude is called "skepticism." Scientists insist on being convinced. To overcome their disbelief, you must present facts.

Surprises: Real science involves trying to uncover mysteries. Because scientists explore the unknown, they often encounter facts that they did not expect. They may even not like what they find out. But because they value new knowledge, successful scientists continue to explore their subjects.

Theory: A theory is a detailed explanation of how something works. For example, long ago most scientists accepted the theory that the sun and the planets move around the earth. This theory explained some observations, but not others. Then a scientist named Copernicus put forth a different theory. He wrote that the earth and the other planets move around the sun. This theory did a better job of explaining the facts, and soon most scientists accepted it.

Vocabulary: Scientists develop and use their own words to describe things. For example, what most people call the "skin" is known as the "epidermis" by skin scientists. When you first study a branch of science, you may find the words strange. But with a little effort, you will master these new words. Then, you'll be able to use them when exchanging ideas with other scientists.

ONGOING PROJECTS

Projects that continue throughout the school year can provide balance for "one-shot" activities.

Agricultural Station: Throughout the year, encourage students to grow things in the classroom. Set a terrarium near a window or use a grow light. Projects might involve growing such things as:
- beans
- vegetables
- herbs
- grasses
- mosses
- wildflowers

Book Reports: Scientists not only immerse themselves in the literature of their fields, they often review books and articles for professional and lay publications. The bibliography, page 95, illustrates the range of excellent books available for young people. On a regular basis, you might have students present formal book reports. Or have each child read books in three or four fields during the course of the year. For a dramatic twist, have a small group read and report on a single book as a panel.

Guest Scientists: Invite scientists to visit your room and discuss their work. Try tapping experts from universities, hospitals, drug companies, observatories, or other institutions. You might also invite outstanding high school scientists.

Online Linking: If you have a Web site at school or at home, collect science links. Invite students and parents to add to the list. We've posted a variety at our Web site and will add more: www.mondaymorningbooks.com.

Question Priming: Reading about science isn't doing science. But to avoid reinventing the wheel, scientists learn to track down facts presented by other scientists. To give students practice in library research, give them interesting questions to answer. You'll find a list on page 93.

Read-aloud Science: Newspapers often feature up-to-the-minute breakthroughs in science. Try reading aloud an article a week. Or if the articles are written at your students' level, have students take turns reporting on current events in science.

Science Fairs: In the upper grades, science fairs can be complex and even competitive events. But smaller, more informal fairs may be more appropriate at the elementary level. Several times a year, students might set up displays in the multipurpose room to share their discoveries with each other. Prizes aren't essential. The goal should be to create a school-wide interest in learning about a wide range of phenomena.

Science in the Public Interest: Encourage students to use their growing scientific knowledge to serve the community in terms of health and safety. Students might create posters, flyers, and Web pages on topics, such as:
- Fire and toy safety
- Dealing with the flu
- Avoiding frostbite
- Memory and study
- Pet care and training
- Diet and exercise
- Poisonous mushrooms
- Recycling
- Sunburn danger
- Product testing

Student Teaching:
A powerful way to master scientific knowledge is to teach it to others. Students can visit earlier grades, giving short lectures and demonstrations. Proven topics include:
- mirrors
- battery-powered electric circuits
- crystals
- magnets
- gyroscopes

Student Writing: Like professionals, young scientists need to share their discoveries and questions. Some outlets:
- SCIENCE BULLETIN BOARD: Students explain developments read about in newspapers and magazines.
- CHILDREN'S BOOKS: Students publish simple science books: describing things (lunar craters), explaining phenomena (how vision works), and teaching methods (how to sprout seeds).
- ARTICLES: Students write reports for the school paper or Web site about events such as the winter solstice or the next eclipse.
- CORRESPONDENCE: Encourage students to pen pal with other young scientists. Some professional scientists, who have put up their own home pages on the Internet, will respond to short and well-written e-mail messages.

Weather Station: Set up a weather station so that students can keep track of the weather on a daily basis. Depending on the level of sophistication of your students, your station might consist of the following items:
- a log for recording data
- an outdoor thermometer
- a barometer
- a rain gauge
- a wind gauge
- a cloud guide

Research Questions

Air: How did scientists find out what air is made of?
Bacteria: How many bacteria live on or inside human beings?
Bees: How do bees tell each other where to find nectar?
Biking: Why do you have to lean when turning a bicycle?
Bread dough: Why does dough rise?
Bubbles: Why do bubbles form in water left standing?
Cold feet: When people get cold, why do their feet get cold?
Cooking: What happens when food is cooked?
Dogs: Why do dogs bark?
Dreams: Why do people dream?
Eyeglasses: How do eyeglasses improve a person's vision?
Flies: Why are flies so difficult to catch?
Flight: Why don't planes have to flap their wings as birds do?
Gesundheit: Why do people sneeze?
H_2O: How did scientists figure out what water is made of?
Hair: Why does hair stand on end when someone is scared?
Hemispheres: Why do the hemispheres have opposite seasons?
Hiccups: What causes and cures hiccups?
Hunger pangs: What makes you feel hungry?
Language: How do babies learn to talk?
Magnifying glass: Why does it flip things if held at arm's length?
Moon: Why does the moon's shape seem to change?
Music: Why do musical instruments sound different?
Nature: Are there any straight lines (or rectangles) in nature?
Pool: Why does cold water feel warmer after a few minutes?
Screeech: Why is the sound of chalk on a board so unpleasant?
Sleep: What makes people get sleepy?
Smoking: How does smoking cause health problems?
Snowflakes: Why do snowflakes have six sides?
Sound: Why can you hear sound from around a corner?
Spheres: Why are the planets and the stars round, not square?
Stars: Why do stars twinkle?
Static: What causes static on a radio?
Stomach: Why does the stomach growl?
TV: How does television work?
Washing: How does detergent help make clothes clean?
Water: Why does water go down the drain in a spiral?

WEB SITES

The Web is rich in science information. If you haven't explored it, the following list can help you get started. It's a multi-way street. We welcome hearing about science-related sites that you discover.

BRAIN: http://uta.marymt.edu~psychol/brain.html
Recipient of a "Los Angeles Times Pick" award, this site presents a digital model that offers many views of the brain.

COMPOSTING: http://www.oldgrowth.org/compost/home.html
This "rotweb" site offers tips from around the world on composting, gardening, and more.

CONSTELLATION QUIZ: http://www.mtwilson.edu/Education/conQuiz/
At this site sponsored by the famed Mt. Wilson Observatory, you'll find a multiple choice quiz on 15 constellations, plus background information.

DINOSAURS: http://www.ucmp.berkeley.edu/diapsids/archosy.htm.
This virtual museum of ancient bones is organized around a map of the sort used by paleontologists to understand how animals are related.

NASA: http://www.hq.nasa.gov/office/pao/Library/photo.html
This site features most of NASA's still images. Topics include astronomy, flight vehicles, mission patches, robotics, and space shuttle photos.

NEWTON: http:/wwwcn.cern.ch~mchab/n/
This site includes facts on Newton's life, an inventory of his estate, quotations from his work, and links to museums featuring his work.

OPTICAL ILLUSIONS: http://www.illusionworks.com/default.htm
A "hall of illusions" features impossible figures, science projects, interactive puzzles, a research library, and links to related sites.

RAINBOWS: http://www.unidata.ucar.edu/staff/blynds/rnbw.html
Everything you might want to know about these spectacular light shows, including diagrams, explanations of double rainbows, and Descartes' description of how he studied the rainbow phenomenon.

SCIENCE LEARNING NETWORK: http://www.sln.org/
Here's the place to find learning projects and museums, on topics ranging from the Amazon to a whirligig farm.

BIBLIOGRAPHY

Many scientists trace their interest in science to childhood reading. Fortunately, today a number of scientists and science writers are turning out remarkably good books on a wide range of science subjects. The following list merely hints at the richness of this important literature.

Astronomy
The Night Sky Book by Jamie Jobb (Little Brown, 1977)
Star Guide by Franklyn Branley (Crowell, 1987)
Starry Messenger by Peter Sis (Farrar, Straus & Giroux)

Biology
Backyard: One Small Square by Donald Silver (Freeman, 1993)
How Did We Find out About Germs? by Isaac Asimov (Walker, 1974)
Lots of Rot by Vicki Cobb (Lippincott, 1981)

Chemistry
Janice VanCleave's Molecules by J. VanCleave (Wiley, 1993)

Ecology
Cartons, Cans, and Orange Peels: Where does your garbage go? by Joanna Foster (Clarion, 1991)

Human Biology
The Body Book by Sara Stein (Workman, 1992)
The Digestive System by Dr. Alvin Silverman (Holt, 1994)
Fingerprint Detective by Robert Millimaki (Lippincott, 1973)
Looking at the Senses by David Suzuki (Wiley, 1991)
The Skeleton and Movement by Steve Parker (Watts, 1989)

Meteorology
Making and Using Your Own Weather Station by Beulah Tannenbaum (Watts, 1989)

Physics
Experiments with Light and Mirrors by Robert Gardner (Enslow, 1995)
How Did We Find Out About the Speed of Light? by Isaac Asimov (Walker, 1986)
Mirrors: Finding Out About the Properties of Light by Bernie Zubrowski (Morrow, 1992)

Psychology
ESP by Daniel Cohen (Messner, 1986)
How to Really Fool Yourself: Illusions for All Your Senses by Vicki Cobb
Theater of the Night: What We Do and Do Not Know about Dreams by S. Carl Hirsch (Rand McNally, 1976).
Memory: How It Works and How to Improve It by Roy Gallant (Four Winds, 1980)

Science Biography
Black Scientists by Lisa Yount (Facts on File, 1991)
Famous Firsts in Medicine by Bette Crook (Putnam's, 1974).
Louis Pasteur by Beverly Birch (Gareth Stevens, 1989)

INDEX

Accuracy, 80
Alphabet, 52
Animals, 6, 7, 72, 73
Anthropology, 46
Apparent motion, 8, 9
Aristotle, 62
Artificial intelligence, 64, 65
Associations, mental, 40, 41
Astronomy, 16, 17, 36, 37
Background knowledge, 86
Balance, 42
Banjo, 44, 45
Biology, 6, 7, 18, 19, 72, 73
Boats, 26
Body, 24, 25, 30, 31, 48-51, 58, 62, 66
Book reports, 91
Chemistry, 68
Classifying, 62, 72, 73
Color, 48
Computer, 64, 65
Conduction, 35
Constellations, 12-15
Craters, 36, 37
Cuts, 30, 31
Cycles, 16, 17
Decay, 18, 19
Decibels, 33
Decomposition, 18, 19
Drawing, 74, 75, 86
Energy, 34, 35, 70
ESP, 20, 21
Evaporation, 22, 23, 58
Experimenting, 76, 77, 86
Eye, 48, 49
Fact checking, 78, 79
Falling bodies, 10, 11
Family tree, 28, 29
Fingerprints, 24, 25
Flip books, 8, 9
Food, 60
Friction, 34

Galileo, 11
Garbage, 18, 19
Genes, 28
Gravity, 10, 11
Healing, 30, 31
Hearing, 32, 33
Heat, 34, 35
Heat stroke, 59
Heredity, 28, 29
Human biology, 20, 24, 28, 30, 32, 46, 50, 52, 58, 60, 62, 64
Hypothesizing, 66, 87
Ice, 68, 69
Inertia, 42, 43
Insulation, 35
Internet, 91
Jargon, 90
Journal, 90
Laws of motion, 43
Lunar craters, 36, 37
Mapping, 60, 61
Measuring, 80, 81
Memory, 38, 39, 77
Mind mapping, 40, 41
Mind reading, 20, 21
Mistakes, 87
Model making, 82, 83
Moon, 16, 36, 37
Motion, 42, 43, 70
Motion pictures, 8, 9
Music, 44, 45
Newton, 42
Notebook, 90
Observing, 6, 18, 46, 52, 62, 66, 84, 85
Optical illusion, 8
Patterns, 16, 17
People watching, 46, 47
Perception, 8
Phonetics, 52, 53
Physics, 10, 22, 26, 34, 44, 54, 68, 70
Planets, 83

Play about healing, 31
Polling, 56, 57
Prediction, 88
Psychic phenomena, 20
Psychology, 38, 40, 56
Public service, 92
Questions, 89
Reading in science, 91
Refraction, 68
Remembering, 38, 77
Reporting, 76, 77, 89
Research, 78, 79
Rot, 18, 19
Safety and science, 89
Scale models, 82, 83
Short-term memory, 38
Senses, 32, 48, 60, 66
Sight, 48, 49
Simulation, 36, 64, 65
Skeleton, 50, 51
Skin, 30, 31
Smell, 60,
Solar still, 23
Speed, 54, 55, 80, 81
Spelling, 77
Superstitions, 56, 57
Stars, 12-15
Survey, 56, 57
Sweat, 58, 59
Taste, 60, 61
Teeth, 62, 63
Temperature, 58, 59
Theory, 90
Thinking, 64, 65
Tongue, 60, 61
Vibrations, 44, 45, 70
Vision, 48, 49
Vocabulary, 90,
Waves, 70, 71
Water, 22, 23, 26, 68
Webbing, 40
Word association, 40
Web, 91, 92, 94